COMPUTER SCIENCE, TECHNOLOGY AND APPLICATIONS

E-WASTE

MANAGEMENT, TYPES AND CHALLENGES

COMPUTER SCIENCE, TECHNOLOGY AND APPLICATIONS

Additional books in this series can be found on Nova's website under the Series tab.

Additional e-books in this series can be found on Nova's website under the e-book tab.

ENVIRONMENTAL REMEDIATION TECHNOLOGIES, REGULATIONS AND SAFETY

Additional books in this series can be found on Nova's website under the Series tab.

Additional e-books in this series can be found on Nova's website under the e-book tab.

COMPUTER SCIENCE, TECHNOLOGY AND APPLICATIONS

E-WASTE

MANAGEMENT, TYPES AND CHALLENGES

YUAN CHUN LI
AND
BANCI LIAN WANG
EDITORS

nova publishers

Nova Science Publishers, Inc.
New York

For permission to use material from this book please contact us:
Telephone 631-231-7269; Fax 631-231-8175
Web Site: http://www.novapublishers.com

LIBRARY OF CONGRESS CATALOGING-IN-PUBLICATION DATA

E-waste : management, types, and challenges / editors, Yuan Chun Li and Banci Lian Wang.
p. cm.
Includes index.
ISBN 978-1-61942-217-9 (hardcover)
1. Electronic waste--Management. 2. Electronic apparatus and appliances--Environmental aspects. 3. Refuse and refuse disposal. I. Li, Yuan Chun. II. Wang, Banci Lian.
TD799.85.E386 2011
621.3815028'6--dc23

2011043397

Published by Nova Science Publishers, Inc. † New York

CONTENTS

PREFACE

With advancements in the electronic world almost occurring on a day-to-day basis and increased availability of products to the public, it is not surprising to see a staggering increase in the generation of electronic wastes over the past decade. This book explores the management, types and challenges related to e-waste. Topics discussed include the processing of plastic materials from e-waste; challenges and opportunities of e-waste management in developing countries; an e-waste quantification study of the People's Republic of China; e-waste environmental contamination and public health effects; and recovery of valuable metals from flat panel displays of spent liquid crystal and plasma televisions.

Chapter 1 - During the last years some research has been running on the development of different methods capable to give added value to wastes from electrical and electronic equipment (WEEE). The rapid growth of WEEE can be attributed to the speed of development in technology, which often has resulted in a reduction of product life-time to less than 2 years, for both domestic and commercial products. Polymers represent approximately a 20% of the total weight of WEEE and include some 15 different types, such as polycarbonate (PC), copolymer of acrylonitrile-butadiene-styrene (ABS), high impact polystyrene (HIPS), polyamides (PA), polypropylene (PP), polyethylene (PE), polyesters etc. In most of the cases, plastic residues generated after disposal of a certain item, mainly engineering polymers, still show good properties. Although, a variety of techniques have been developed for recycling polymers, in general and particularly for WEEE, the high cost associated with these methods usually leads to the disposal of plastics to landfills. The main drawback that obstructs material recovery from plastic contained in WEEE is the variety of polymers that have being used, resulting to a difficulty in sorting and recycling. Another relevant drawback is that very often they contain brominated aromatic compounds, used as fire retardants. Thermal treatment of such chemicals is likely to produce extremely toxic halogenated dibenzodioxins and dibenzofurans. Other barriers to recycling these plastic components include difficulties with dismantling, removal of impurities, such as rubber feet, labels etc. Insufficient knowledge regarding the compatibility of these materials during melt processing and service life of the final product makes recycling a rather complicated problem. In this chapter the recent developments on the recycling and recovery of plastic wastes from WEEE will be reviewed, with special emphasis on each type of the above mentioned materials, as well as on the recycling process and the potential reuse. The four routes of plastic waste management treatment will be discussed covering primary (re-extrusion), secondary (mechanical), tertiary (chemical) and quaternary (energy recovery) schemes and technologies. Experimental results related with the structure and properties of recycled ABS, PC, HIPS and PP were also presented, with special emphasis on the effect of melt re-processing on the performance of binary mixtures among them. This latest approach might be interesting since these polyblends, when properly compatibilized, often exhibit improved properties, beyond those of their components. In addition, this process would be of importance to the industrial practice, as it can give final products with acceptable properties by direct mixing of the as received plastic waste stream, i.e. without prior separation and sorting.

Chapter 2 - In this review, the global production of e-wastes is described based on recent surveys and statistics. The variety and projections in e-waste composition is also identified due to the technological improvement in the electronic industry, which introduces difficulty for the development of a universal, sustainable recovery process. The effects of various organic and metal components of e-wastes on human health and the environment are summarized based on the literature findings. This review includes the challenges related to the rapidly increasing e-waste generation and describes the dilema that the developing countries are facing and the opportunities brought along with e-wastes to these countries. The chapter also includes a proposed process for the treatment and recocery of Printed Circuit Boards (PCBs).

With advancements in the electronic world almost occurring on a day-to-day basis and increased availability of products to the public, it is not surprising to see a staggering increase in the generation of electronic wastes over the past several decades. According to a United Nations Environment Program (UNEP) alert bulletin in 2006, in the manufacturing industry of industrialized countries, the production of electrical and electronic devices is the fastest-growing sector of industrial development. Thus the handling and treatment of electronic waste, or so-called E-waste, is a topic of concern in the future. Globally, there is no doubt that the amount of E-waste being produced and in turn disposed is sharply increasing. Although it is hard to give an accurate estimate of the global E-waste production due to faulty and sometimes non-existent data, the UN estimate of the global E-Waste production in the same alert was 20-50 million tons per year. In 1998, the E-waste production in the United States was estimated at 6 million tons per year. A more recent report released by the UNEP in July 2009 has shown stunning global facts indicating rapidly growing generation of e-waste.

The adverse environmental and health risks caused by E-Waste are of growing importance. E-wastes are made of a variety of organic substances and metals in a polymer matrix. The organic substances, mainly fire retardants and other additives are potential carcinogens. They accumulate in the food chain and the environment due to their non-biodegradable nature. Toxic heavy metals such as cadmium, chromium, mercury and lead are released to the environment mainly through leaching and incineration of the e-wastes. These metals are highly toxic and pose high risk to human health and are a major environment impact. In addition to the toxicity of the e-waste materials, the production process of electronics is often a very energy intensive endeavor. For example, the production of a single personal computer and monitor takes at least 240 kg of fossil fuels, 22 kg of chemicals, and 1.5 tons of water. This amount of water alone, is more than the weight of a modern automobile. This situation further increases the burden to the environment.

Having realized the hazards of E-waste, many developed countries are taking actions. In January 2003, The European Union Council addressed the serious issue of electronic waste streams by the Waste Electrical and Electronic Equipment (WEEE) Directive to motivate the reuse, recycling and reduction of WEEE by efficient material recovery methods. It also seeks to minimize the environmental impacts of WEEE by conducting life cycle assessments. Another regulation, the Restriction of Hazardous Substances (RoHS) Directive, has also become EU law in February 2003. It has been established to restrict the use of hazardous substances in WEEE aiming to reduce its impacts on human health as well as the environment. Followed by these EU's pioneer actions, other countries such as Canada, the United States, and the UK, have developed (or partially developed) e-waste related laws and directives in recent years.

Although the developed countries attempted to reduce the amount of WEEE by regulation, huge quantities of their WEEE are still exported to developing countries including China, Pakistan, India and Indonesia through different channels. The low concentrations of precious metals in e-wastes provide enough economic incentive for less affluent countries to attempt resource recovery methods. However, most of their e-waste treatment facilities and metal recovery processes are primitive which create significant pollution to the air, water and land resources. The local people in turn suffer from a variety of syndromes due to over-exposure to the toxic substances. Thus, the technological advances in the separation and purification of metals and even organic substances from e-wastes are crucial for minimizing the effects on the environment in the recovery processes.

To tackle these challenges from the increasing production and disposal of e-waste, more actions should be taken by the governments including arousing the public awareness, developing environmentally friendly separation and purification processes for recovering valuable materials from e-wastes, establishing new environmental policy and laws, providing safety measures for workers in the recovery facilities, etc. In all probability, e-waste will sooner or later become a commodity due to the abundance of limiting resources such as gold, silver, indium and other precious metals in these solid wastes.

What the authors suggest is the concept of "mining from the e-wastes". With advanced separation technologies, the processes can be developed to take full advantage of e-wastes in an environmentally friendly manner. Here, a sustainable and feasible treatment process for printed circuit boards is proposed and described. Its applicability for metal recovery from e-waste is discussed.

Chapter 3 - Digital divide refers to the gap between those who have access to technology and those who do not. Increasing quantities of used electronic products are exported to some developing countries annually; a greater proportion being non-functional and un-usable. Consequently, large quantities of electronic waste (e-waste) are being managed in developing countries using inappropriate management methods including: disposal at open dumps, open burning and the use of crude extraction techniques in recovering valuables. Reasons given for the trans-boundary movement of e-waste appears plausible and laudable - it will benefit those who are too poor to afford brand new equipment thereby *bridging the digital divide*, result in environmental gains as cheaper labour in the developing countries can make repair of old equipment feasible, giving it longer life, and even recycle e-waste. The dumping of e-waste in developing countries is unsustainable, environmentally unjust and undermines the health of the ecosystem. This practice exposes developing countries to an array of highly toxic components of e-waste. The digital divide can be bridged in sustainable ways that provide digital opportunity for the poor without worsening their condition. This can be achieved by assisting developing countries with 'tested' and "functional" "used electronic devices" and not 'electronic scrap' to be recycled. There is therefore the need for a global policy initiative and legislation in dealing with this emerging global crisis.

Chapter 4 - With increasing exposure of the harm from electronic and electrical waste (e-waste) recycling in the People's Republic of China, it is an imperative that its generation and recycling be sufficiently regulated. Accurately depicting the rate of generation and disposal of e-waste is an important first step in planning for adequate formal e-waste logistics and treatment capacities.

A critical review of the quantification studies on China's e-waste generation and disposal rates shows that there are large discrepancies in the predicted amounts of e-waste generated no matter the same or different methods of estimation were used. Moreover, even when agreements between studies are found, the agreed figures are not necessarily the correct figures. There appears to be an imminent need to increase the transparency of data collection and computation methods in these studies as the lack of such transparency is identified as the main reason for estimation uncertainty.

Evident from this chapter, irresponsible citation practices are found to spread to academic studies already. It is urged that leading organizations in the academic community should consider establishing a platform devoting to the reporting of false or dubious citation.

Chapter 5 - Electronic waste (e-waste), which contains hazardous material, is much more expensive and difficult to dispose and recycle than other kinds of common garbage. E-waste has raised many concerns worldwidely due to its rapid generation; trans-boundary shipment from developed countries; improper recycling in developing countries; environmental impacts and health hazards. Property tags showed that US, Japan and UK are major exporters of these wastes to Guiyu, a town 40 km from the Shantou, Guangdong province, China. Local environment matrix, such as air, soil, water, sediment and leachate, has been found containing high concentration of hazardous material routed from e-waste recycling process. With extremely primitive and uncontrolled methods, nearly 6,000 family workshops in Guiyu process approximately 1.6 million tons of e-waste yearly and produce extensive pollutants which have adverse effects threatening local community. Majority of these workers are also residents of these recycling sites thereby increasing their exposure rates. The authors' studies found that people employed in e-waste recycling in Guiyu had a high incidence of skin damage, headache, vertigo, nausea, chronic gastritis, and gastroduodenal ulcers. Guiyu children and neonates have significantly elevated blood lead, cadmium and chromium levels with impairment of neurobehavioral development and damaged lymphocyte DNA. Guiyu neonates also showed higher rates of adverse birth outcomes such as preterm delivery, stillbirths and low birth weight. The authors' studies suggest that environmental pollution by uncontrolled e-waste process has adversely affected health condition of local population, especially fetus or children health and development, which may cause long-term impairments for them.

Chapter 6 - Recovery processes of indium from waste liquid crystal display panels and that of silver from display panels of spent plasma televisions were developed on the basis of hydrometallurgical technology. In the former case, the powdered sample was leached with 3 M HCl and extracted metals like indium, aluminum, tin and so on were mutually separated and recovered by passing through two columns: the first column was packed with porous resins impregnated with Aliquat 336, a commercially available solvent extraction reagent of quaternary ammonium compound, while the second column was packed with those impregnated with Cyanex 923, also commercially available solvent extraction reagent of trialkylphosphine oxide. In the first column, tin, iron and zinc were removed from the leach liquor while, in the second column, only indium was selectively recovered. The metal ions trapped in these columns were eluted with 0.1 M H_2SO_4 concentrated more than about 10 times.

On the other hand, in the latter case, metals including silver were leached from powdered scrap of plasma TV monitors using acidothiourea solution, which achieved almost complete extraction of silver. The recovery of silver from acidothiourea leach liquor was achieved by means of simple cementation method by using zinc powder. Moreover, precipitation with sodium sulfide was found to be a promising means for silver recovery. Persimmon tannin solution, a natural and nonconventional coagulant, containing polyphenolic groups, was also found to completely recover silver from silver-containing acidothiourea leach liquor through coagulation-precipitaion followed by reduction of silver.

Chapter 7 - Pursuant to the Resource Conservation and Recovery Act (RCRA), the U.S. Environmental Protection Agency (EPA) has established regulations regarding the disposal of hazardous wastes. Although there are federal requirements under RCRA for the management of hazardous waste, some states have opted to implement *more* stringent requirements — particularly with regard to the management of certain hazardous wastes generated by households and small businesses (entities that are essentially exempt from RCRA's hazardous waste management requirements).

One category of household hazardous waste that many states are choosing to regulate more strictly is electronic waste, commonly referred to as "e-waste." E-waste generally refers to obsolete, broken, or irreparable electronic equipment like televisions, computers and computer monitors, laptops, printers, cell phones, copiers, fax machines, stereos, or video gaming systems. Cathode ray tubes (CRTs) in televisions and computer monitors have presented a particular concern to states, primarily due to the potentially significant amounts of lead they contain and the large numbers in which they are generated.

State concerns specific to the landfill disposal or incineration of e-waste are largely due to its increasing volume and often bulky nature; hazardous constituents, such as lead and mercury, it may contain; its high cost of recycling; and the inability of interested stakeholders, such as electronics retailers and manufacturers, to reach consensus on how to voluntarily implement a national e-waste management system. States have responded to this concern by enacting their own e-waste management laws. Requirements of those laws range from a ban only on the landfill disposal or incineration of designated e-wastes to the implementation of a full e-waste collection, transportation, and recycling system.

To date, 13 states have enacted some form of e-waste management law (as many as 20 states proposed e-waste laws in 2006 and 2007). Although the goal of each law is similar — to avoid landfill disposal and incineration of certain types of e-waste —approaches taken to achieve that goal differ significantly. However, most state laws and proposals have certain broad elements in common, such as specifying the electronic devices covered under the law; how a collection and recycling program will be financed; collection and recycling criteria that must be met to minimize the impact to human health and the environment; and restrictions or requirements that products must meet to be sold in the state.

As more states propose e-waste legislation, potentially regulated stakeholders (particularly electronics manufacturers and retailers) have expressed concern that they will be required to comply with a patchwork of state requirements throughout the United States. This concern has led to an increased call for federal legislation regarding e-waste management. To help policy makers better understand the impact of state e-waste legislation, this report discusses issues that have led to state action, common elements in state-waste laws and proposals, and an overview of each enacted state law.

Chapter 8 - Expansion in electrical and electronic equipment trade has led to significant increase in electronic waste which should be dealt with special priority due to its potential negative impact to the public health and the environment. Thermal plasma technology offers a very promising alternative of hazardous waste treatment for the near future. Plasma treatment is advantageous where landfill is difficult or impossible, where storage costs are encountered, or where stringent emission standards make combustion-based treatment expensive. This is usually the case for hazardous materials, in particular for e-wastes which contain heavy metals.

Plasma technology is an effective solution to treat the most toxic wastes like organohalogen wastes, contaminated hospital wastes, fly ashes, radioactive wastes, and e-wastes. Plasma torch technology has been developed to produce temperatures as high as 4000 ° C. At these temperatures, any substance is vitrified, i.e., reduced in volume and transformed into a glassy obsidian whose leachability is many times less than bottle glass. Thermal plasma technology takes the advantages to incinerate the combustible parts of wastes for volume reduction and to vitrify the noncombustible counter parts simultaneously into glassy slags with very low leaching rate. Furthermore, and compared with other waste treatment technologies, the principal advantages of plasma technology are the reduction of exhaust gas flow rates, in most cases higher volume reduction, installations with smaller footprints and lower investment costs for a given throughput, and faster start-up and shut-down times.

In this chapter, the authors will define and give the different types and common forms of plasma. The authors will also present an overview of plasma processes for waste treatment, and give a review of solid waste treatment by plasma. In section 5 the authors will present the technique of hydrogen recovery from the thermal plasma gasification of solid wastes. However in section 6, the authors will present the long- term behavior of vitreous or crystallized materials obtained by vitrification of solid wastes. In the last section they will present a study on the e-waste treatment by plasma.

Chapter 9 - Electronic wastes have become the foremost stream of solid waste in almost all the countries of the world over a short span of last few years. The digital information and communication technology (ICT) revolution that the mankind is undergoing has substantially contributed in creating this scenario. Popularly abbreviated as e-wastes, electronic waste comprises of all out-of-date discarded or otherwise unused electronics and peripherals thereof – the bulk being computers of many descriptions, consumer electronics and mobile phones. Necessarily, a huge amount of electronics is replaced by individuals and organizations each year but people are increasingly purchasing more electronics than they really need with the rapid introduction of cheaper, sleeker and alluring gadgets. The technology and trends are changing quite rapidly rendering usable electronics obsolete and make us to go for newer versions of the same thing thereby adding up to the volume of outmode electronics. Bulk of these ends up into landfills, for USA 80-85% of e-wastes goes to landfills as estimated by US-EPA. Volume of e-wastes is a big concern but the high content of heavy metals, *e.g.* lead, chromium, cadmium and mercury, etc., along with harmful organic compounds in the electronic products is a bigger concern for mankind. To some, the name 'e-waste' is ironic as it contains over 60% recyclables including iron, copper, aluminum, gold and other metals let alone the polymers from casings of disposed electronic products. The increasing flow of e-wastes naturally stimulated various solution responses, one of which is the boom in e-waste recycling business, in the form of formal industry in developed part of the world while manual and archaic disassembly of wastes in developing countries most of which are imported from developed countries for recycling. The non-recyclable parts of these wastes, including the toxicants, are finding their way into the environment and it has led to the formulation of Basel Ban, an amendment to the Basel Convention to restrict trans-boundary movement of e-waste. However, the e-waste flux from developed to developing countries is on the rise due to win-win economics of informal recycling. The authors' knowledge about the chemistry of distribution, speciation, transport and environmental fate of these hazardous elements in and outside landfills are still sketchy though some researchers have initiated in-depth studies to assess and predict the eco-environmental consequences of the toxicants from disposed e-wastes. The incineration at the informal reprocessing sites contaminates the air and soils with polychlorinated and polybrominated dibenzo-*p*-dioxins and dibenzofurans, polycyclic aromatic hydrocarbons, etc. On the other hand, acid leaching processes contaminate the freshwater sources and sediments with heavy metal. The effects of these releases are exacerbated through the entry of pollutants into the food cycle. Besides the contamination of the surroundings, the recycling process itself poses health risks to the workers out of exposure to the chemicals used in recycling and hazardous byproducts from the recycled wastes. Within the scope of this chapter, the authors will provide a brief overview, in light of the recent findings, of the toxic environmental releases from the e-waste disposal and recycling processes and their effects.

In: E-Waste: Management, Types and Challenges
Editors: Yuan Chun Li and Banci Lian Wang
ISBN 978-1-61942-217-9

Chapter 1

PLASTIC MATERIALS FROM E-WASTE: CLASSIFICATION, PROCESSING AND REUSE

Marianna I. Triantou and Petroula A. Tarantili
National Technical University of Athens
School of Chemical Engineering, Polymer Technology Lab.
Zographou, Athens, Greece

ABSTRACT

During the last years some research has been running on the development of different methods capable to give added value to wastes from electrical and electronic equipment (WEEE). The rapid growth of WEEE can be attributed to the speed of development in technology, which often has resulted in a reduction of product life-time to less than 2 years, for both domestic and commercial products. Polymers represent approximately a 20% of the total weight of WEEE and include some 15 different types, such as polycarbonate (PC), copolymer of acrylonitrile-butadiene-styrene (ABS), high impact polystyrene (HIPS), polyamides (PA), polypropylene (PP), polyethylene (PE), polyesters etc. In most of the cases, plastic residues generated after disposal of a certain item, mainly engineering polymers, still show good properties. Although, a variety of techniques have been developed for recycling polymers, in general and particularly for WEEE, the high cost associated with these methods usually leads to the disposal of plastics to landfills. The main drawback that obstructs material recovery from plastic contained in WEEE is the variety of polymers that have being used, resulting to a difficulty in sorting and recycling. Another relevant drawback is that very often they contain brominated aromatic compounds, used as fire retardants. Thermal treatment of such chemicals is likely to produce extremely toxic halogenated dibenzodioxins and dibenzofurans. Other barriers to recycling these plastic components include difficulties with dismantling, removal of impurities, such as rubber feet, labels etc. Insufficient knowledge regarding the compatibility of these materials during melt processing and service life of the final product makes recycling a rather complicated problem. In this chapter the recent developments on the recycling and recovery of plastic wastes from WEEE will be reviewed, with special emphasis on each type of the above mentioned materials, as well as on the recycling process and the potential reuse. The four routes of plastic waste management treatment will be discussed covering primary (re-extrusion), secondary (mechanical), tertiary (chemical) and quaternary (energy recovery) schemes and technologies. Experimental results related with the structure and properties of recycled ABS, PC, HIPS and PP were also presented, with special emphasis on the effect of melt re-processing on the performance of binary mixtures among them. This latest approach might be interesting since these polyblends, when properly compatibilized, often exhibit improved properties, beyond those of their components. In addition, this process would be of importance to the industrial practice, as it can give final products with acceptable properties by direct mixing of the as received plastic waste stream, i.e. without prior separation and sorting.

A. Characteristics and Processing of Plastics from E-Waste

1. Plastic Waste Management

Plastics are low-cost, lightweight and durable materials, which can readily be moulded into a variety of products that find use in a wide range of applications. After their end of life, many plastics gather in landfill's spaces. Landfilling exhibits many drawbacks, such as the contamination of soils and groundwater by some additives and breakdown by-products in plastics, which can become persistent organic pollutants, the reduction of spaces for landfill and mainly the lack of recovery of the material resources used to produce the plastics. Incineration reduces the need for landfilling of plastic wastes. However, there are concerns that hazardous substances may be released into the atmosphere in the process [1]. Re-use and recycling of plastic wastes are considered alternative routes of waste management. Re-use of plastic items, which end up in the waste stream after a single use only (single-life or cycle) or a short time after purchase, is always noted as a preferable choice to recycling, as it uses less energy and fewer resources, conserves fossil fuels and reduces carbon-dioxide (CO_2), nitrogen-oxides (NO_x) and sulphur-dioxide (SO_2) emissions [2].

Recovery and recycling of plastic solid wastes (PSW) can be categorized by four main routes, i.e. re-extrusion, mechanical, chemical and energy recovery [2]. In the recycling industry, sorting occurs by both automatic and manual methods. Automated sorting involves density sorting, triboelectric separation and speed accelerator techniques. Another main issue is the removal of the paint on plastics. Grinding, abrasion, solvent stripping and high temperature aqueous-based method are used for this purpose [3].

Re-extrusion (primary or closed-loop recycling) involves re-introducing scrap, industrial or single-polymer plastic edges and parts to the extrusion cycle in order to produce products of similar material [3]. This process uses scrap plastics that have similar features to the original products [2].

When plastic articles are discarded after a number of life cycles, mechanical (secondary or downgrading) recycling techniques present themselves as a candidate for utilizing a percentage of the waste as recyclate and/or fillers. In particular, mechanical recycling is the process in which PSW are used in the manufacturing of the plastic products via mechanical means [2]. Thermoplastics, including poly(ethylene terephthalate) (PET), polyethylene (PE) and polypropylene (PP) all have high potential to be mechanically recycled. Thermosetting polymers, such as unsaturated polyesters or epoxy resins, cannot be mechanically recycled.

This is because those plastics are permanently cross-linked in manufacture and therefore cannot be re-melted and re-formed [1]. One of the main issues in mechanical recycling is the degradation and heterogeneity of PSW. Since chemical reactions that constitute polymer formation (i.e. addition, condensation, etc) are all reversible in theory, energy or heat supply can cause photo-oxidation and/or mechanical stresses which occur as a consequence. Length or branching of polymer chains and/or cross-linking can also occur from the formation of oxidized compounds and/or harsh natural weathering conditions. Recycling PSW mechanically involves a number of treatments and preparation steps to be considered, which are size reduction (by milling, grinding or shredding), separation and washing. In the case of polymer foams powdering takes place in the preparation stages [2]. The most general scheme of mechanical recycling includes cutting/shredding, contaminant separation, floating, milling, washing and drying, agglutination, extrusion and quenching [3].

All technologies that convert polymers to either monomers (monomer recycling) or useful petrochemicals (feedstock recycling) are referred to as chemical (tertiary) recycling. Under the category of chemical recycling advanced process appear, e.g. pyrolysis, gasification, liquid-gas hydrogenation, viscosity breaking, steam or catalytic cracking and reduction in blast furnaces, hydrolysis, glycolysis, fractionation, hydroglycolysis, aminolysis, methanolysis and acid cleavage. The main advantage of chemical recycling is the possibility of treating heterogeneous and contaminated polymers with limited use of pre-treatment [2]. Biodegradable plastics can also be composted, and this is a further example of tertiary recycling, described as organic or biological recycling [1].

Energy (quaternary) recovery employs combustion processes to produce heat, steam and/or electricity. Since the calorific value of plastics is high, they make a convenient energy source [2]. The main process units of a waste-to-energy plant are: waste feed system, combustion section and boiler, heat recovery system, exhaust gas cleaning system and ash handling system [4]. The presence of flame retardants complicates the technical aspects of energy recovering [3]. PSW combustion generates CO_2, NO_x, SO_x, volatile organic compounds (VOCs), smoke, particulate–bound heavy metals, polycyclic aromatic hydrocarbons (PAHs), polychlorinated dibenzofurans (PCDFs) and dioxins [2]. CO_2 along with water, the principal products of fuel combustion, can influence global warming. Other environmental impacts include destruction of the ozone layer, acid rain, resource depletion, energy demand and soil and water pollution. Monitoring of halogen and volatile metal content in the fuel is necessary to avoid overriding the gas cleaning system, corrosion problems and compromising clinker or ash quality in cement industries that can negatively impact disposal or beneficial use options [4]. Capture and removal of flue gases in thermal and combustion processes is a major issue dealt with by (i) ammonia addition to combustion chamber, (ii) flue gas cooling, (iii) acid neutralization, (iv) activated carbon addition and/or (v) filtration [3]. Energy recovery processes may be the most suitable way for dealing with highly mixed plastic, such as some electronic and electrical wastes and automotive shredder residue [1].

Feedstock recycling technologies are more costly than mechanical recycling and less energetically favourable as the polymer has to be depolymerized and then re-polymerized. Recycling has been shown to save more energy than that produced by energy recovery, even when including the energy used to collect, transport and re-process the plastic [1].

In 2002, 20.6 M tonnes of available plastics at Europe were managed as follows: 14% to mechanical recycling, 2% to feedstock recycling, 23% to energy recovery and 62% to landfill and incineration without energy recovery. The electrical and electronic plastics were managed: 96% incineration without energy recovery and landfilling, 4% mechanical recycling and <1% energy recovery [4].

2. Plastics from Electrical and Electronic Equipment Waste (WEEE)

2.1. Characteristics of WEEE Stream

It is well known that electrical and electronic (E&E) products continue to revolutionize communication, entertainment, transportation, education and health care around the world. The demand for EEE is continuously growing, thanks to the market expansion, the development of new types of equipment and the increasingly rapid replacement of old technologies with novel ones. As a result of the continuous change in the design and function of consumer E&E products, the mechanical properties of the obsolete products, i.e. WEEE, are highly variable. The above differences can be explained by the number of different appliances and the heterogeneity in composition of any given appliance [5].

Information technology (IT) represents today the fastest manufacturing waste producer. It is estimated that there are over a billion personal computers in the world. In developed countries, these have an average life span of only two years.

In Europe, Japan and United States, the E&E market makes up approximately 88 percent of the overall consumption of plastics, which is well behind packaging and building and construction products, but comparable to automotive [4]. The high content of bromine, chloride and heavy metals in WEEE has led to the need for separate collection and recycling scheme for this type of waste, in order to reduce environmental impacts. The joint disposal of WEEE and municipal solid waste by landfilling or incineration releases hazardous substances and causes resource losses.

Based on the above considerations, during the last years some work has been carried out in the development of different methods to give added value to WEE. A range of techniques are currently applied for retrieving components and materials from WEEE. The essential features of these systems generally conform to a scheme of: sorting/disassembly, size reduction and separation. Most recycling plants utilize manual sorting and disassembly. Manual sorting involves the removal of hazardous components, such as batteries or the sorting into classifications, such as high and low grade material. Disassembly allows the removal of a component, part, group of parts or a sub-assembly from a product (partial disassembly) or the separation of a product into all of its component parts (complete disassembly) for a defined purpose. Size reduction consists of mechanical processes, such as physical impaction, shredding/fragmentation and granulation. Also, a variety of separation methods are available, including screeners, air classifiers, concentrating tables or density separators, electrostatic separators, floatation systems, magnetic separators, rake, spiral and bowl classifiers, trammels and water classifiers [6].

Recycling of WEEE plastics is a big challenge, mainly due to two reasons: First, the plastics waste fraction of WEEE consists of more than 15 different polymer types [7]. The plastics commonly encountered in EEE are Acrylonitrile-Butadiene-Styrene (ABS), Polycarbonate (PC), PC/ABS blends, High Impact PolyStyrene (HIPS) and Polyphenylene Oxide blends (PO) [6]. Second, a large part of WEEE plastics contain brominated flame retardants (BFRs), including polybrominated biphenyls (PBB) and polybrominated diphenylether (PBDE) [7]. 30 percent of all E&E plastics contain flame retardants and 12 percent of all plastics used in E&E sector contain brominated flame retardants. The main product categories that use flame retardants are television housings, computer monitor housings, CPUs and circuit boards. Some plastics parts on copiers and printers, especially laser jet printers can be flame retarded due to the potential for high temperature spikes. Plastics used in industrial electrical equipment often contain flame retardants due to the high voltages involved [4].

2.2. Insights of Recycling and Re-Use of Plastics from WEEE

The main difficulty of handling WEEE in the industrial level is the separation of different materials present in the mixture. One of the most successful approaches of separation systems is based on the physicochemical properties of materials to make recycling of materials constituting WEEE economically profitable. Before proposing a recycling procedure for WEEE, a previous complete characterization of each individual component is necessary to get all the information on the differences in mechanical, chemical and thermal properties between materials in WEEE and original polymers and blends.

Taurino et al. [7] identified the presence of ABS in televisions and ABS/PC blends in personal computers using Fourier transformed infrared spectroscopy (FTIR), elemental analysis and differential scanning calorimetry (DSC). Also, they established reliable detection of BFRs, especially of tetrabromobisphenol A (TBBPA) and decabromodiphenyl ether (DeBDE), with a combination of techniques suitable only for waste deriving from personals computers. Several impurities, like TiO_2 pigment, are easy to identify for both devices, i.e. televisions and personal computers [7].

Schlummer et al. [8] utilized density separation, which allows the separation of styrenics from other polymers present in the WEEE stream (e.g. PE, EP, UP, POM, PC). However, the corresponding density window overlaps those of polyamide (PA), glass-fibre-reinforced or talc-filled polypropylene (PP-GF, PP-T), rigid polyurethane (PU) or with plasticized polyvinylchloride (PVC). Consequently, density-based polymer separation alone does not permit pure recycled polymers to be obtained. For this reason CreaSolv® process, a solvent-based technology, has been used. This process allows the removal of non-polymeric materials such as dust, metals and glass splinters as well as unwanted additives and contaminants, such as brominated flame retardants (BFR), polybrominated dioxins and furans (PBDD/F) from the polymer solution [8].

Dimitrakakis et al. found that ABS is the largest fraction in samples taken from of small size WEEE (small household appliances e.g. toasters and shavers, information technology and telecommunications equipment e.g. telephones, mobile phones and calculators, consumer equipment e.g. radios and speakers, toys, leisure and sports equipment e.g. videogame consoles), followed by PP, PS, PC and PVC [9]. About half of the plastics contain brominated flame retardants, while in the majority of the cases Sb_2O_3 is also used as a synergistic additive. Most elements are found in concentrations below 0.1% (w/w), but concentrations of Br, Sb, Zn, Cu, Fe and Ti reach or even exceed 10% (w/w).

Flame retardants are utilized for reasons of fire safety. The use of flame retardants delays the spread of fire or the time of flashover. The choice of the recommended flame retardant will depend on several criteria, such as the application, resin, fire safety standards that must be met, cost considerations and recyclability [10]. Halogenated FR have traditionally been used because of their efficiency and suitability with various plastics. Bromine is generally preferred over chlorine because it requires lower quantities of additive and minimizes the impact of its incorporation on the polymer's performance. In particular, BFRs are incorporated into plastics either though reaction or addition. Reactive combination produces plastics with covalently bonded flame retardants. These flame retardants are less likely to leave the product. In contrast, additive flame retardants are simply mixed with the polymer resin and so continually leach out of the final product. Moreover, the thermal treatment of such chemicals is likely to produce extremely toxic halogenated dibenzodioxins and dibenzofurans [7]. Environmentally friendly fire retardant systems currently developed to substitute halogen based systems will decrease the content of halogen in forthcoming WEEE, however actual WEEE and those collected in near future still contain large amounts of brominated FR.

Elimination of BFR polymers by sorting appears to be reasonable. However, manual bromine and polymer identification at dismantling lines is time and man-power consuming and, hence, automated sorting systems are preferred, which allow the treatment of shredded materials. Since current high-throughput spectroscopic sorting systems are restricted to particle sizes larger than typical shredder material and cannot treat black polymers, non-spectroscopic means are required for shredder separation at present. Density-based technologies were also applied as sorting technologies since BFR-polymers exhibit higher densities compared to corresponding non-BFR types [11].

Other barriers to recycling these plastic components include difficulties with dismantling, removal of impurities (such as rubber feet and labels), insufficient knowledge as to how recycling affects these materials and the viability of mixing various different plastics that are used to make components for EEE. A further problem is the lack of efficient collection or take-back systems for such equipment.

Another economically and environmentally advantageous option is to use plastic waste for energy recovery using thermal treatment, thus reducing the overall use of fossil fuels and the amount of greenhouse gas emissions, while avoiding landfilling. Thermal treatment comprises different processes such as direct combustion, co-combustion with organic wastes, pyrolysis and gasification. In all these processes, a minimal amount of bromine contained in the waste stream turn into ashes or char, while most turns into combustion gases or syngas. Using suitable wet scrubbing systems, it is technically feasible to recover bromine, which can be used to produce different types of commercial bromine-based products [12].

Bientinesi and Petarca [13] investigated the environmental impact of different thermal treatments of WEEE, by the application of a life cycle assessment methodology. Two scenarios were assessed: (i) the co-combustion of WEEE and green waste in municipal solid waste combustion plant and (ii) the staged-gasification of WEEE and combustion of produced syngas in gas turbines. Their results showed that the staged-gasification had a potentially smaller environmental impact than co-combustion. This study demonstrated also that WEEE thermal treatment for energy and matter recovery is an eco-efficient way to dispose of this kind of waste.

In the pyrolysis recycling strategy of WEEE, the gas fraction should be suitable for HBr recovery, the oil for chemicals recuperation or fuel production, the residue for precious metal recovery. Luda et al. studied the pyrolysis of flame retarded high impact PS and epoxy resins by both flow and closed systems [14]. The effect of alkali on pyrolysis was also studied, demonstrating, as far as epoxy resin is concerned, a efficiency on decreasing bromine content in oil and volatile products, further leading to the recovery of bromine from the residue by washing. The alkali treatment was shown to be less effective in styrenic polymers containing brominated flame retardants [14].

Marco et al. [15] pyrolyzed polyethylene (PE) wires, table and mobile phones as well as printed circuit boards (PCB). They concluded that the solids remaining after pyrolysis are composed of metals and other inorganics present in the raw material, mixed with a char-like carbonaceous material formed during pyrolysis. Metals can be easily separated and recovered and char may find application as pigment or component of asphalt fabrics. As regards pyrolysis liquids, they vary significantly depending on the type of the pyrolysed WEEE stream. For example, liquids from PE wires are mainly composed of long-chain non aromatic compounds and consequently have a wax-like, semi solid consistency at room temperature and, therefore, are difficult to handle. The liquids of both, phones and PCB, are mainly aromatic, of low viscosity and easy to handle. However, they contain significant amounts of nitrogen and other heteroatoms-containing compounds, so that in order to use such liquids as fuels, blending with petroleum refinery streams and/or upgrading should be needed. Alternatively, PCB and mobile phones pyrolysis liquids may be used as source of chemicals, such as phenol or styrene. WEEE pyrolysis gases are mainly composed of hydrocarbons C1 to C5 together with significant proportions of CO and CO_2. Concerning hydrocarbons, ethylene/ethane are the predominant ones in PE and PCB gases, while methane is the basic constituent in table and mobile phones gases [15]. As the collection of end-of-life (EOL) E&E products from households and business increases around the world, the search for economically, environmentally responsible and sustainable recovery and recycling intensifies. Whenever plastics are unsuitable for mechanical recycling or other routes of material recovery on eco-efficiency grounds, then the use of plastics for energy production, as an alternative to landfilling, needs to be thoroughly analyzed from the point of view of sustainable development. Capturing the heat value of E&E plastics (comparable to oil or natural gas) as an alternative to the use of fossil fuels can be a key component of integrated resource management and energy policy, generating hot water, steam and electricity. In modern energy recovery systems, capturing the fuel value of plastics that cannot be recycled, might be among the cleanest forms of energy. Recovering metal from the ash and beneficial use of the ash provide additional value to the process [4]. The gasification recycling technology of plastics WEEE containing brominated flame retardants has also been examined. After high temperature treatment (1200°C), shock cooling of gases to a temperature of roughly under 200°C was done by the proposed process [16]. The generated "thin" gas could then be used as raw materials for chemical industries or for electric power generation. The effect of the high temperature treatment and the shock cooling suppresses the emission of brominated dioxins to a very low level, just as for chlorinated dioxins.

3. Basic Plastic Components on WEEE

3.1. Poly(Acrylonitrile–Butadiene–Styrene) Terpolymer (ABS)

3.1.1. Recycling of ABS

ABS is an amorphous thermoplastic composed by acrylonitrile (between 20 and 30%), styrene (20-60%) and an elastomeric phase butadiene (20-30%) [17]. The consumption of

ABS-ASA-SAN in electrical and electronic equipment was 496.000 tonnes in 2000. ABS, HIPS and PUR foam are the predominate plastics found in refrigerators and freezers [4].

Recycling of ABS wastes from electric and electronic equipment (WEEE) premises the removal of metallic particles and glass. While magnetic separation has been successfully employed for eliminating the ferrous particles, the recycling industry does not dispose yet of a satisfactory solution for the non-ferrous particles. Iuga et al. established that electrostatic and corona-electrostatic separation methods are suitable for removing the metallic particles that contaminate the ABS products recycled from wastes [18].

Another important problem of ABS recycling is the difficulty to classify plastics into each material [19]. Brennan et al. examined samples that were taken from disused plastic casings of computer equipment and concluded that ABS is the most widely used plastic and HIPS is the second most common plastic [20].

The above authors also found that poly(phenylene oxide) (PPO), PC/ABS and ABS/PVC blends are used in moderate proportions, whereas only a small amount of casings are made of acrylic, PVC, PP and PMMA [20]. Park et al. [21] developed a reliable and relatively inexpensive technology for the effective triboelectrostatic separation of PVC, PET, and ABS by a two-stage separation process, based on the control of the charge polarity and charge-to-mass ratio of those particles. Particularly, their study was focused on estimating process variables, such as air velocity, electrode potential, splitter position and relative humidity that affect the separation efficiency of the plastics [21].

Achilias et al. recovered ABS from computer monitors, TV sets and electronic toys using the dissolution/reprecipitation method [22]. It is a simple and economic method leading to high recovery of pure polymer and can be optimized as regard the dissolution temperature and time, polymer concentration, type and portion of solvent/non solvent. For model ABS, high recoveries are measured when using dichloromethane/methanol at 100°C, as well as acetone/methanol at lower temperatures. Concerning the system acetone/methanol, although it results in good recoveries when tested with model ABS, it is not effective in WEEEs based on ABS. For all different WEEEs, based on ABS the best solvent was found to be dichloromethane at 100°C with recoveries greater than 90 wt %. No significant changes in the chemical structure of the polymer recovered with this technique can be identified by FTIR spectrums [22].

The superiority of recycling is that CO_2 emissions of recycled ABS are smaller than those of the virgin material at any rate of recycled ABS. The reason is that CO_2 emissions in the recycling process are smaller than those corresponding to the manufacturing of virgin polymer, starting from mining of raw materials to the polymerization process [19].

Polymer recycling and reuse of recycled polymers for the design of new product components is steadily increasing. It is accepted that direct use of post-consumer polymers is the most efficient way to treat plastic waste. However, the performance of products made of recycled polymers will differ from products made of virgin grades, because the initial processing step creates degradation mechanisms and several reactions occur, that can be detrimental to the properties of polymer, including oxidation, cross-linking and decomposition.

3.1.2. Structure and Properties of Recycled Products

The decrease of physical properties of recycled ABS, when it is mixed and when platings and paintings cannot be separated from the plastic body is, as well, an important issue.

Therefore, it is required that all information about the base materials and additives, such as flame retardants, is visible on the surface. This will allow to classify them, to use single or highly miscible materials and to avoid decorations such as plating, painting and labeling or to choose the decorations which have high miscibility with the resin [19]. Garcia et al. investigated the compatibility of recycled PVC/ABS blends and found partial compatibility due to the interaction between the polar groups of the SAN phase and those of PVC [23]. Low molecular weight species, which exist in the PVC, migrate towards the SAN and thereby, create a SAN-rich phase with certain amounts of PVC dissolved within it. Dissolved PVC acts as an internal plasticizing agent. Similarly, the migration of the low molecular weight species from PVC towards SAN dominions generates a PVC-rich phase but with reduced mobility. The partial compatibility of two components is confirmed by the variation of glass transitions temperature (T_g) as a function of the blend concentration. However, the use of virgin ABS offers higher compatibility than the recycled ABS, because of the improved adhesion which exists between the polybutadiene phase and the SAN phase. For compositions with high concentration of ABS, the PVC blend with recycled ABS shows a greater concentration of ABS dissolved in the PVC, due to the presence of a greater number of low molecular weight chains, as a result of possible degradation produced by its service life or during reprocessing. PVC admits greater amounts of ABS, both virgin and recycled, than vice versa [23].

Brennan et al. studied the recycling of ABS and HIPS from waste computer equipment and found that blending proportions of these materials has no significant effect on T_g, tensile strength and tensile moduli [20]. The presence of 10% ABS in HIPS and vice versa has a favorable effect on the strain to failure, providing some recovery of the significant amounts of ductility that are generally lost in the individual polymers due to degradation during recycling. However, as recycling has an adverse effect on the impact properties of ABS and HIPS, due to both polymer degradation and the presence of pigment TiO_2 in each polymer, causing some inhomogeneity, blending of HIPS into ABS decreases the impact strength even further. The addition of ABS in HIPS does not affect impact properties further than those due to recycling.

Mechanical recycling is the main recovery option for managing ABS from end-of-life electronics. The degradation of plastics materials in conventional use can be regarded as the sum of all types of weathering effects occurring during the product's life cycle, both degradation in manufacturing processes and weathering during the product's use. In order to understand the changes taking place on the properties during recycling, it is relevant to study the effects of combined processing and weathering in several repeated cycles. Pérez et al. investigated the reprocessing of ABS using extrusion and injection and the effect of combined conventional extrusion or injection processing and UV weathering on ABS [24]. Boldizar and Möller studied the effect of combined conventional extrusion processing and ageing in air at 90°C for 72 h on the properties of a conventional ABS-grade [25]. Bai et al. simulated recycling cycles using a torque rheometer and injection molding [26]. Arostegui et al. [27] studied a dissolution-based recycling technique for ABS using acetone as solvent with 0.25 g/ml concentration, at room temperature and 40 min for dissolution. The above authors also studied the effect of repeated recycling cycles [27].

Polymer recycling and reuse of recycled polymers for the design of new product components is steadily increasing. It is accepted that direct use of post-consumer polymers is the most efficient way to treat plastic waste. However, the performance of products made of recycled polymers will differ from products made of virgin grades, because the initial processing step creates degradation mechanisms and several reactions occur, that can be detrimental to the properties of polymer, including oxidation, cross-linking and decomposition. The above mentioned reactions of oxidative degradation and cross-linking obviously affect the molecular weight and molecular weight distribution of the material. An additional difficulty is that returned polymers have usually been exposed to a variety of thermal and mechanical environments.

One alternative capable of maintaining the properties of the final product to an acceptable level is to blend virgin material with recycled feedstock. The effect of varying composition of a recycled ABS resin in two types of virgin ABS, GPPC-D350 and Toray-100MPM, on the physical and mechanical properties as well as on the molding characteristics of the final blend was investigated by Chen et al. [28]. By increasing the weight percentage of the recycled material, the melt flow index and T_g of blend increases slightly. There is no significant effect of content of recycled ABS on the tensile and flexural properties, but the impact and torsion strength of blend decreases with increasing recycled resin. The opposite tendency is recorded for hardness. The injection pressure decreases, as the concentration of recycled ABS increases, due to the reduced melt viscosity [28].

3.1.3. Effect of Recycling Cycles on Thermal Properties

Many researches [24-26, 29] report that multiple reprocessing has no apparent effect on the T_g of the SAN phase. Pérez et al. [24] concluded that T_g of polybutadiene (PB) phase and styrene-acrylonitrile (SAN) phase in the extruded and injected ABS are not affected, as well, by its chain length and SAN phase's composition. The T_g of weathered samples does not depend either on the number of reprocessing cycles or on the residence time in the QUV chamber [24]. Boldizar et al. [25] observed that the T_g of SAN phase is almost constant for the samples from combined and extrusion-only series, as opposed to ageing-only series where the T_g increases, possibly due to physical ageing. Bai et al. [26] found that the T_g of PB increases after 4 cycles of treatment, probably due to crosslinking and thermo-oxidative degradation taking place in the rubber phase. Arostegui et al. [27] confirmed the degradation of polybutadiene phase with FTIR analysis, where it is observed a slight decrease at the height of absorption bands associated with the butadiene double bonds and a new absorption band related to saturated carbonyl groups for dissolved ABS and dissolved and injected ABS in comparison with injected ABS. The appearance of this new absorption band is in agreement with the fact that degradation of butadiene takes place especially on the unsaturated bonds, leading to a new product with carbonyl groups. Moreover, the above authors reported that the degradation of ABS takes place also in the SAN phase, which can primarily be attributed to the injection process rather than to the dissolution process [27].

With respect to the initial degradation temperature, it is practically kept constant up to 10 reprocessing cycles by both injection and extrusion techniques, as mentioned by Perez et al. [24]. The thermal stability increases slightly as a function of the residence time in the chamber, probably due to elimination of monomers and/or additives caused by the photooxidation processes [24]. Boldizar et al. [25] observed a small decrease in the reduction of the oxidation temperature through the extrusion series, a somewhat larger reduction in the ageing series and an even greater decrease in the combined series. Arostegui et al. [27] suggest that during dissolution some low molecular weight compounds, such as heat stabilizers or antioxidants, are probably eliminated and consequently degradation takes place mainly in the following injection moulding step. Karahaliou and Tarantili [29] observed a decrease of char residue of ABS from the first to the fifth extrusion cycle, which could be attributed to chain scission taking place during the extrusion processing and, furthermore, to the higher oxygen content of the processed ABS samples, as indicated by the surface darkening of the samples with the progress of successive extrusion cycles. This change was measured by colour testing and, moreover FTIR spectroscopy was run in order to explore the formation of oxidation products, such as esters, hydroxyls, carboxyls and carbonyls progressively through the repeated cycles [29].

3.1.4. Effect of Recycling Cycles on Mechanical Properties

Pérez et al. [24] observed that the tensile properties, which are controlled by the rigid SAN phase, are unaffected by the number of reprocessing cycles. However, the tensile strength of weathered materials decreases, whereas the modulus of elasticity remains constant. Upon exposure to UV radiation, photolysis of the methylene bond in the trans-1,4-polybutadiene structure occurs, giving rise to an allylic radical polymer chain. This radical may initiate several reaction paths to yield further radicals, together with oxygenated species such as hydroperoxides, ketones and esters. As a consequence, crosslinking reactions can take place in the polybutadiene phase causing higher fragility which results in a lower tensile strength. For the samples after 10 reprocessing cycles, the property is reduced by 50%, whereas for the samples after the fifth reprocessing cycle the decrease is only 20%. This behavior can be explained taking into account that the photooxidation mechanism is accelerated by the presence of residual hydroperoxides formed during the processing, which may also decompose to produce other degradation products and radical chains. The impact strength decreases significantly after recycling, probably as a consequence of some degradation of the PB phase involving changes in the molecular weight [24].

Decrease in impact strength after multiple reprocessing was noticed by Bai et al. [26]. The greatest reduction is during first cycle, as the major loss in volatile molecules found during this cycle. Farther reduction at following cycles is attributed to crosslinking of the rubber phase and scission of the ABS polymer chains. Furthermore, they observed that the modulus of ABS do not change significantly with increasing number of reprocessing cycles, but tensile strength increases slightly [26]. No apparent effect on tensile strength and modulus, but increase in elongation at break in third and fourth cycle and reduction in next cycle is observed by Karahaliou and Tarantili [29].

Boldizar and Möller [25] discovered that elongation at break increases in the extrusion series and decreased in the ageing series. In the combined extrusion and ageing series, elongation at break remains fairly constant during the first cycle, whereas in the second to the sixth cycles increases during each extrusion step and decreases during each ageing step. This behavior is associated with both physical and thermo-oxidative degradation. Due to less restricted oxygen diffusion, the thermo-oxidative degradation of the material is expected to proceed more rapidly at the surface than in the bulk of the sample. For the extrusion-only series, it is therefore reasonable to expect less accumulation of degradation products in the surface than in the case of the ageing series, due to redistribution of the products as a result of the mixing action of the extrusion process. In the combined and ageing series, the loss of unsaturation and the increase in carbonyl index may produce a brittle surface layer, more prone to initiate cracks into the bulk of the material [25]. Arostegui et al. [27] found out that low-strain properties (Young's modulus) are not affected by the number of recycling cycles, but high strain properties, as yield stress and impact strength, are reduced due to the butadiene degradation and SAN's lower molecular weight. These properties drop in the first recycling step and remained more or less constant in the successive recycling cycles [27].

3.1.5. Effect of Recycling Cycles on Rheological Properties

No significant changes were observed by Pérez et al. [24] for flow index upon increasing the reprocessing cycles. The melt flow index (MFI) from the first to fifth extrusion cycle remains almost constant, according to Karahaliou and Tarantili [29]. Similarly, Boldizar and Möller have found that the melt volume-flow rate (MVR) remains almost constant in the extrusion as well as in the ageing series [25]. Also, in the combined series, the MVR values remains fairly constant up to the fourth cycle, but increases during the following cycles, maybe due to the consumption of antioxidant stabilizers during the extrusion, which results in a less protected ABS against the thermo-oxidative ageing process [25].

3.2. Polycarbonate (PC)

3.2.1. Recycling Methods

Polycarbonate is an amorphous thermoplastic constituted by long linear chains of polyesters of carbonic acid and phenol, such as bisphenol A [17]. Bisphenol A polycarbonate exhibits superior dimensional stability, good electrical properties, good thermal stability and outstanding impact strength [30]. The consumption of PC in electrical and electronic equipment was 53.000 tonnes in 2000. PC, PPO/HIPS, PP and PC/ABS appear in relative large parts in consumer electronics, IT equipment and small household appliances [4].

Achilias et al. recovered PC from compact disks using the dissolution/reprecipitation method [22]. For model PC, high recoveries were recorded using dichloromethane/methanol, even at low temperatures. The recovery of PC from compact disks is high enough (98%) in accordance with results of the model polymer. Moreover, in the same work pyrolysis of waste compact disks and model polycarbonate was studied as recycling process. After pyrolysis at 550°C, a large amount of oil is formed, together with a smaller amount of gaseous products, leaving also a solid residue. For both samples, the gaseous fraction consists mainly of CO and CO_2, whereas in the liquid fraction a large amount of different phenolic compounds including the monomeric bisphenol A, was detected [22].

Chiu et al. [31] concluded that the use of metal chlorides $SnCl_2$ and $ZnCl_2$ as catalysts in the pyrolysis will be beneficial and efficient for recycling treatment of PC as both metal chlorides, in particular $SnCl_2$, can effectively lower the degradation temperature, shorten the degradation time and narrow the liquid product distribution in comparison to the pyrolysis of PC alone. The optimum metal chloride loading in the degradation reaction depends on reaction conditions [31].

Watanabe et al. [32] investigated the chemical recycling of polycarbonate at high pressure and high temperature steam. For 5 min reaction time, polycarbonate completely decomposes into bisphenol A (monomer of PC) and the maximum yield of bisphenol A is around 80%. The high yield of bisphenol A in high pressure steam is due to the high stability of bisphenol A. PC decomposition at 573 K in liquid water phase was also studied for comparison. For 30 min in reaction, residual PC still remained and the bisphenol A yield is about 50% as maximum. The amount of water required for polycarbonate degradation was drastically reduced and, thus, the high pressure-high temperature steam process was energetically and economically preferable [32]. Liu et al. [33] reposted that PC can be methanolysized to bisphenol A (BPA) and dimethyl carbonate (DMC) in the presence of ionic liquid [Bmim][Cl] without any acid or base catalyst under moderate conditions. Under the conditions of m(PC):m(CH_3OH):m([Bmim][Cl])=2:3:2, reaction temperature 105°C and time 2.5 h, the methanolysis conversion of PC is almost 100% and the yields of both BPA and DMC are over 95%. The ionic liquid could be reused 8 times without obvious decrease in the conversion of PC and yields of BPA and DMC [33].

Endo et al. [34] studied the synthesis and chemical recycling of a polycarbonate obtained by anionic ring-opening polymerization of a bifunctional cyclic carbonate consisting of both five- and six- membered rings (65CCP). The depolymerization of P(65CCP) takes place at 60°C for 24 h with 1,8-diazabicyclo[5.4.0]-7-undecene (DBU) as a catalyst. The original polymers are completely transformed to oligomers and 65CCP by the nucleophilic attack of the imine nitrogen in DBU to the carbonyl group in the main chain to produce alkoxide. The amount of regenerated 65CCP and M_n of the resulting product reach a constant value after 15 h, and 65CCP is recovered in 86% yield. The high recovery ratio suggests that P(65CCP) depolymerizes in an unzipping manner, which is similar to common depolymerization of vinyl-type polymers. The saturation of the depolymerization ratio is ascribable to the equilibrium nature of this polymerization-depolymerization system [34].

3.2.2. The Effect on Properties and the Role of Additives

Mechanical recycling of engineering polycarbonate wastes provides an interesting way to decrease their impact on environment [35]. The combined effect of reprocessing and accelerated ageing on the mechanical and thermal properties of polycarbonate was studied by Pérez et al. [36]. Up to 10 reprocessing cycles have been carried out by both injection and extrusion techniques. The above authors concluded that reprocessing causes a decrease in the molecular weight of the PC and, therefore, a decrease in the T_g and the initial degradation temperature (T_i). This behaviour may be the consequence of the chain scission reactions originated in the thermal and shear degradation processes that take place during reprocessing. The accelerated ageing causes a decrease in the molecular weight of the polycarbonate, especially during the first ageing period, independently of the number of reprocessing cycles. Decrease of the T_g of recycled PC after eight injection cycles and an increase of the storage modulus from the first cycle to the fifth (then remains constant up to the eighth cycle) were observed by Feller and Bourmaud [35]. This effect can be associated to the decrease of molecular weight taking place during processing.

No significant difference in the T_g after six injection mouldings is noted by Zenkiewicz et al. [37]. There is not much change in the thermal stability of the PC potassium diphenylsulfone sulfonate (KSS) after eight recycling cycles, as reported by Statler et al. [30]. However, the colour turns to yellow and then to brown due to a slight thermal degradation of the polymer. KSS is a good flame retardant additive for a recycled polycarbonate and this was demonstrated by improving the flammability properties of a PC sample coming from unknown sources and containing a mixture of different flame retardants [30].

With respect to mechanical properties, according to Pérez et al. [36] the modulus of elasticity or the tensile strength are not affected from the first to seventh reprocessing cycle, whereas the impact strength decreases sharply. However, studies based on accelerated ageing show an important influence of the number of reprocessing cycles on mechanical properties after the first reprocessing cycle [36]. Statler et al. [30] found that the Young's modulus, yield stress, ultimate stress and elongation at break of PC flame retarded with either bromine or potassium diphenylsulfone sulfonate stay comparatively constant [30]. On the other hand, the Izod impact strength suffers the most during recycling. After the first and second injection mouldings, minor increases (ca. 2.2 wt.% in sum) in the tensile strength of polycarbonate were observed by Zenkiewicz et al. [37]. After the next injection mouldings, this quantity does not change much.

Decrease of viscosity with number of cycles was observed by many researchers [30, 35, 37]. This phenomenon is attributed to the decrease of molecular weight and increases with decreasing temperature [35]. Drozdov et al. [38] compared the viscoelastic behaviour of melts of virgin PC (VPC), recycled PC (RPC]) or their mixture (MPC) reinforced with sort glass fibers (SGFs) and evaluated the effect of recycling of PC on time-dependent behaviour of the composites with various amounts of filler. Three series of shear oscillatory tests were performed at 250 and 290°C. Constitutive equations were derived for the viscoelastic behavior of polymer melt at isothermal deformations with small strains. A polymer was treated as an equivalent transient network of strands that rearrange at random times as they agitated by thermal fluctuations. For VPC, the instantaneous shear modulus G increases with temperature. At 250°C, the elastic modulus of VPC increases with the content of glass fibers (v), whereas at 290°C, decreases with v. For RPC and MPC, the elastic modulus decreases with the content of glass fibers v. The rate of rearrangement, γ, increases with temperature and decreases with the content of fibers, due to constrains on mobility of polymer chains imposed by the presence of short glass fibers. The rate of rearrangement γ for virgin PC exceeds that for recycled PC, which is explained by the contributions of short chains in the recycled material that serve as extra physical cross-links in the transient network. The average energy for rearrangement of strands, V, decreases with the content of fibers, because the strength of entanglements between chains noticeably exceeds the strength of interactions between macromolecules and glass fibers. The rate of decrease of V for recycled PC exceeds that for virgin PC, which is associated with the better adhesion of chains to short glass fibers in virgin PC than in recycled polymer [38].

Sun et al. [39] used two types of core-shell structured latexes, poly(methyl methacrylte-co-butadiene-co-styrene) (MBS) and poly(methyl methacrylate-co-methylphenyl siloxaneco-styrene) (MSiS) to modify recycled polycarbonate (PC) in terms of the enhancement of toughness and flame retardancy. The impact strength of the modified PC blends is not improved after melt-blending of recycled PC with these two kinds of latexes, probably because the latex particles are not evenly dispersed in the PC matrix because of the incompatibility between PC and PMMA shell of the latexes. Addition of a compatibilizer, e.g. diglycidyl ether of bisphenol-A or poly(styrene-co-maleic anhydride), can effectively enhance the toughening effect of recycled PC with core-shell structured modifiers. The ternary blends achieve a high impact resistance by cavitation of the particles, which relieves the triaxial stress and promotes massive shear yielding of the matrix and then enables the matrix to fracture by the plane stress ductile tearing mode. Additionally, MSiS has a silicone-based core and can effectively retard the combustion of recycled PC. The blends containing 7 wt % MSiS and 3 wt % compatibilizer can achieve a UL94 V-0 rating in vertical burning test. The good flame retardancy obtained is attributed to the fine dispersion of the MSiS particles in the PC matrix, which facilitates the rapid migration of MSiS and formation of a uniform and highly flame resistant char barrier on the surface of the modified PC [39].

Nishihara et al. [40] studied the moldability and recyclability of silicone containing PC resins. The linear D-type silicone is superior to branched DT-type silicones and other conventional agents in its flame retardant effect in PC resin because of its high silicone content, intra-resin mobility and PC stabilization. The flame retardance is attributed to the action silicon and not the silicone itself. It permits the use of PC with relatively low molecular weight with no loss in impact strength and, thereby, improves melt flowability in the moulding process and it is ecofriendly, in terms of safety and recyclability. These characteristics make it highly suitable for office equipment, computers and other applications requiring high heat resistance and performance [40].

3.3. ABS/PC Blends

Polymer blends produced by the mixing of ABS and PC allow the preparation of products with a wide range of physical, chemical and mechanical properties. In particular, blending PC with ABS decreases the cost and improves processability of PC by reducing melt viscosity and enhancing shear-thinning behavior and also increases the heat distortion temperature of ABS material. There are a variety of forms (dispersed, continuous or co-continuous) in which the rubber (SAN-grafted polybutadiene) and the plastics (SAN and added PC) may take, depending on the composition, processing conditions, compatibility and multiphase interactions [41]. The morphology of the mixture is a determining factor on the mechanical properties of the blend. Waste polycarbonate can be effectively toughened by maleic anhydride grafted ABS (ABS-g-MA). It was found that in this system, the effect of maleic anhydride grafting onto ABS is not an improvement of compatibilization between two kinds of segments of ABS, but the generation of special morphology of ABS domains. First, different segments of ABS are combined in an enwrapped structure, which partially eliminates the repulsion between the polar (SAN) segments and the non-polar PC ones. Second, at a certain PC/ABS-g-MA weight ratio, the dispersed ABS domains connected together forming a network, which gives rise to a maximum impact strength [42].

Other factors that can play an important role on the mechanical and rheological properties of the blends are the degradation taking place in the earlier stages and the miscibility characteristics of their components. The presence of an elastomeric phase, which is especially sensitive to thermo-oxidative degradation, is one of the most influencing factors of the final performance, since its degradation promotes a decrease on adherence with SAN matrix acting as stress concentrators. Butadiene phase does not take part in miscibility phenomena because of its network structure [43]. Eguiazábal and Nazábal [44] observed that the small deformation properties, such as modulus of elasticity or yield stress do not change with reprocessing. As far as the high-strain mechanical properties are concerned, such as tensile strength, ductility and impact strength, only a slight change after one or two processing cycles can be observed followed by a significant decrease after more than two processing cycles. Reprocessing also affects the density and MFI values, where clear changes appeared between the behavior before the second processing cycle and that displayed after this cycle [44].

Cândido et al. studied the effect of recycling cycles of materials (RCM), by testing a 70/30 ABS/PC blend [17]. The RCM was determined by five stages of reprocessing in a twin-screw extruder. They observed an inappreciable decrease in tensile properties and a slight increase in the impact strength of the blends with RCM, whereas the impact strength is inferior with respect to the virgin ABS and virgin ABS/virgin PC blend.

A decrease in mechanical properties, in terms of tensile strength and elongation, is recorded by Tarantili et al. for pure PC/ABS mixtures [45]. However, the modulus of elasticity shows some improvement, due to the improvement of stiffness of the corresponding blends. ABS seems to act as plasticizer when incorporated into PC at concentrations of about 20%, probably due to the dispersion of the polybutadiene phase within the thermoplastic matrix of the blend. This effect was also clear during the determination of rheological characteristics of the PC/ABS blend via MFI measurements. Here again, the sample with composition 80/20 shows increased flow rate which suggests easier processing [45].

PC/ABS blend comprising 40 wt% recycled PC, 10 wt% low molecular weight virgin PC and 50 wt% virgin ABS would be an acceptable product having a high recycle content and balanced processing and mechanical properties, as it is mentioned by Liang and Gupta [41]. Jin et al. studied ABS/PC blends and reported that as the content of PC increases the impact strength and Vicat softening temperature (VST) were also increased [46]. As the melt viscosity of ABS increases and takes values near to that of PC, more efficient distribution of dispersed PC phase and consequent enhanced impact strength and VST were observed.

Balart et al. [43] found that the composition range between 10 and 20 wt% of recycled PC in recycled ABS is the most interesting in order to obtain an industrial material with balanced properties for following reasons: (i) mechanical ductile properties do not decrease, (ii) processing conditions are similar to those for other styrenic derivatives and (iii) this composition range reflects the generation ratio of these wastes, which is close to 4:1 for ABS/PC.

For safety reasons, plastics used in electronic and electrical components contain flame retardant materials in order to conform to various fire safety regulations [47]. Dawson and Landry [10] studied the multi-pass (5 injection molding cycles) of PC/ABS blends contained diphenyl phosphate (RDP) and bisphenol A diphenyl phosphate (BPADP) as flame retardant. They concluded that PC/ABS samples contained BPADP have superior recyclability than that contained RDP, as they maintain the properties better. The same conclusion arises for PC/ABS samples, which underlie in extended heat and humidity. Potential reasons include the flame retardant type, type and ratio of PC/ABS used and other additives used in the formulations [10].

Therefore, if recycled blends are to be used in electronic applications they must also possess acceptable flammability properties. Khan et al. [47] discovered that the addition of virgin or recycled PC to virgin or recycled ABS appears to progressively decrease extinguishment time, burning extent and weight loss on fire. The addition of 70% of PC (virgin or recycled) to ABS (virgin or recycled), appears to significantly decrease the heat and smoke release. The results of this study indicate that recycled polycarbonate can be used as an additive for virgin or recycled ABS, as a means of providing flame resistance to ABS in high-value applications [47].

3.4. High Impact Polystyrene (HIPS)

3.4.1. Recycling and Stability

Another very interesting material, also found in WEEE, is high impact polystyrene (HIPS). This polymer is widely used, especially in those products where brittle polystyrene (PS) cannot meet the required standards of impact resistance. HIPS can be structurally defined as a multiphase system that exhibits a continuous rigid phase together with dispersed rubber particles, which vary from 0.5 to 10 μm in diameter. The rubbers commonly used in the HIPS synthesis are polybutadiene or styrene/butadiene copolymers.

As already mentioned, the thermal treatment associated with recycling could modify the polymer's mechanical and thermal properties and, more specifically in the case of HIPS, is complications may arise because there are two material phases, each of which may react in a different way. Kalfoglou and Chaffey [48] studied the effect of repetitive extrusion processing at high temperatures on the structure parameters and properties of HIPS. It was reported that rheological and tensile properties changed significantly, while the elongation and the impact resistance changed only at high processing temperatures (290°C), decreasing to 57 and 29%, respectively. These changes were attributed to the breakage of the interactions between the matrix and the rubber particles as well as to the rubber degradation, which is a consequence of the number of extrusion cycles and/or to the high temperature employed during processing.

Recyclability of HIPS is not only important in the context of WEEE management but also in thermoforming, where HIPS sheets can give 25-30% scrap, which is to be recycled [49]. HIPS was repeatedly coextruded at 220°C, maintaining a constant composition of 70 wt% of virgin HIPS and 30 wt% of recycled HIPS. The Melt Flow Index decreased in the first cycle, while from the second to sixth increased. The power low index, increased from n=0.29, after the first processing cycle, to n=0.34 in the sixth cycle. The above changes were attributed to changes in the physical structure of the rubber phase and to chain scissions in the PS matrix, caused by the recycling. The impact strength decreased with the increasing number of processing cycles, while the tensile stress at break remained constant.

A simulation of the degradation of HIPS, occurring during service life and mechanical recycling was performed by Vilaplana et al., by multiprocessing and thermo-oxidative ageing [50]. The reprocessing was performed on virgin HIPS, by multiple extrusion up to nine cycles employing a double screw extruder, while the degradation processes occurring during service life was simulated by thermo-oxidative ageing in a forced-ventilation oven at 90°C. Their results showed that HIPS is a promising material for mechanical recycling, since its properties are not extremely affected after multiple processing. However, thermo-oxidative ageing affects more severely the oxidative stability and the mechanical properties of HIPS. Therefore, considering the whole life cycle and the mechanical recycling potential of HIPS material, the previous service life of the material seems to determine in a very important way the degree of degradation of the HIPS recyclates and its further possibilities of employment in second market applications.

Multiple processing affects PB microstructure and the free-volume parameter associated with the PS phase. Physical ageing of the PS phase predominates for shorter exposure to thermo-oxidation; after prolonged exposure, however, the chemical effects on the PB phase become significant and strongly influence the overall structure [51]. The PB phase has been reported to be the initiation point of photo-oxidative, thermo-oxidative, and thermo-mechanical degradation in styrenic copolymers, which induce alterations in the chemical structure of the material [52]. The oxidation of tuber-modified polymers shows an initial induction time, followed by a fast oxidation of PB and then by a slow oxidation rate corresponding to PS degradation.

Parres et al. [53] studied the effect of injection moulding temperature and the application of successive pyrolysis cycles on the butadiene phase of HIPS. Thermal analytical methods did not reveal great changes in the thermal properties of HIPS. Pyrolysis-gas chromatography/mass spectroscopy (Py-GC/MS) - a much more sensitive technique - was also applied for the detection of small structural changes. The results using the above technique did not show significant changes in terms of the injection-molding temperature. They showed differences in butadiene compounds, however styrene compounds do not change with injection moulding temperature. Important changes were observed for pyrolysis cycles in virgin HIPS. Butadiene compounds were very sensitive to temperature, and the results showed that small compounds increase as a function of the pyrolysis cycles.

Despite the research in progress relevant to HIPS recycling, it should be noted that, currently, only 1% is recovered from the total 19% market share of HIPS [54,55].

3.4.2. Study of Mixtures with other Polymers

ABS and HIPS are two of the plastics most frequently used as outer casing for computer equipment such as monitors, keyboards, and other similar components. Brennan et al. [20] assessed the effects of the recycling and blending of ABS and HIPS on mechanical properties. They found that the effects of recycling on ABS and HIPS were similar, in that changes in glass-transition temperatures, tensile strengths and tensile moduli were negligible, but strains to failure and impact strengths were reduced considerably. Blending proportions of ABS and HIPS caused no more deterioration in properties than occurred as a result of the recycling process, and the presence of small proportions of one material in the other actually restored significant amounts of ductility, as seen by increases in the strains of failure. In order to overcome some of the barriers that hinder the process of recycling activities, such as dismantling, removal of impurities, classification and sorting, attempts to recycle mixtures of HIPS and ABS have been made [20].

In fact, it was reported that most thermoplastics may be reprocessed with small amounts of other polymers without substantial decrease in mechanical properties [56]. More specifically, previous work has shown that multiple recycling passes had negligible effect on the tensile properties of the above polymers, although more apparent effects were observed in elongational and impact properties [57-59].

It seemed therefore interesting to run a parallel study on ABS/HIPS systems deriving form melt mixing of the above polymers. The investigation of rheological characteristics along with some structure parameters and final properties of the mixture could provide useful information for their recycling process. In addition, ABS/HIPS mixtures could be suitable for uses as additives in PC recycling, instead of ABS alone, contributing to the low cost of final products. This is reasonable taking into account that the dispersed rubber phase in HIPS could act in a similar mode as the butadiene phase present in ABS, i.e. as shock absorber that greatly contributes to impact resistance of the recycled mixture of plastics.

Campomanes and Manrich studied the compatibilizing effects of a triblock copolymer poly(styrene-b-ethylene-co-butylene-b-styrene) (SEBS) in postconsumer and virgin PP/HIPS blends [60]. Morphological studies provided evidence for the interfacial activity of block copolymer attributable to a stronger interaction of SEBS between the dispersed phase and the matrix, reducing average particle size of the HIPS and increasing interfacial adhesion between both phases.

3.4.3. Property Changes after Recycling

Vilaplana et al. [61] suggested a combination of thermal analysis, vibrational spectroscopy and chromatography techniques in order to monitor the changes in properties of recycled HIPS throughout mechanical recycling processes. The presence of other polymeric materials that may considered as impurities, the structural properties, the oxidative stability, the chemical changes and the presence of low molecular weight compounds in recycled HIPS were some of the key parameters for the quality assessment of recycled HIPS. Vibrational spectroscopy showed the presence of oxidized parts of the polymeric chain gave also an assessment of the microstructure of the PB phase in HIPS. The presence of low molecular weight compounds in the HIPS samples was demonstrated using microwave-assisted extraction followed by gas chromatography spectroscopy (GC-MS). Styrene and styrene dimmers were identified already in virgin PS; in addition benzaldehyde, α-methylbenzenal-dehyde, and acetophenone were detected in recycled HIPS. The presence of oxygenated derivatives of styrene may be attributed to the oxidation of PS [61].

HIPS makes up a significant proportion of WEEE plastics and is often fire retarded with decabromodiphenyl ether and diantimony trioxide to decrease its flammability. As mentioned above, the flame-retardant content of WEEE, particularly the presence of bromine and antimony, represents a major hindrance to the development of recycling technologies. Mitan et al. [62] studied the controlled pyrolysis of PE/PP/PS mixed with brominated HIPS containing decabromodiphenyl ether as a brominated FR with antimony trioxide as a synergist. Analysis with GC with electron capture detector confirmed that the bromine content was rich in step 1 (oil 1) liquid products leaving less bromine content in step 2 (oil 2) liquid products. In the presence of antimony containing samples, the major portion of bromine was observed in the form of antimony bromide and no FR species were found in oil 1. In the presence of synergist, the step 1 and step 2 oils contain both light and heavy compounds. By controlling the pyrolysis parameters and the subsequent pyrolysis steps can produce the liquid hydrocarbons with minimal amount of brominated hydrocarbons.

Accelerated thermal and photo-aging for four homopolymers: LDPE, HDPE, PP and HIPS, was performed and the impact of subsequent reprocessing conditions on their properties was studied by Luzuriaga et al. [63]. In the case of HIPS a slight photo-degradation of PS matrix was accompanied by simultaneous crosslinking of the PB component. In a recent work, the effect of the degree of degradation of the components of a model municipal plastic waste (a mixture of LDPE, HDPE, PP and HIPS) on the toughness and stability of recyclates compatibilized with a cooperative compatibilization system (a mixture of ethylene-propylene statistical and styrene-butadiene block copolymers with a secondary amine-based stabilizer) was studied [64]. It was shown that good impact strength was achieved for recyclates having components with a low or medium degree of degradation. Mechanical properties of recyclates having the components with a high degree of degradation were deteriorated. The addition of the cooperative compatibilization system leads to a higher thermo-oxidative stability of recyclates irrespective of the degree of degradation of their components. Photo-oxidative stability of the recyclates is low but it can be improved to a satisfactory level by the addition of carbon black or a commercial photo stabilizer.

3.5. Polypropylene (PP)

3.5.1. Mechanical Recycling

PP is an important commodity plastic and is used extensively for packaging, medical products, ducts, chemical tanks, etc. However, its contribution to the field electrical and electronic products is appreciable, especially in the market of electric appliances (refrigerators, washing machines) etc. With increasing use, the amount of discarded PP items in the waste stream is also increasing. Management of PP post consumer waste is therefore crucial and, recycling is a viable option for waste management. Recyclability of PP is dependant on its structure and properties.

Reprocessing of PP by multiple extrusion produces a reduction in molar mass and polydispersity, which is more pronounced when the extruder screw has kneading elements [65]. Mechanical recycling of PP is routinely done by milling post-consumer goods, followed by extrusion and pelletization. A part of the stabilizers are consumed during processing, one way to minimize reduction in properties during mechanical recycling is stabilization with incorporation of new additives before reprocessing.

Primary recycling is common in the plastics industry using milled residues free of contamination. Post-consumer recycling is used in applications with lower requirements. For certain applications, the addition of post consumer PP (PPpc) to virgin PP, can lower the costs, reducing raw material consumption and environmental impact, with maintenance of the intended properties. The use of PPpc is feasible, but, depending on its lifetime, this material is partially degraded and contains contaminants, showing inferior mechanical properties in relation to virgin PP [66].

PP was injection moulded up to ten times to mimic the effect of recycling procedures. The main effect of recycling is the lowering of the melt viscosity, which was attributed to molecular weight decrease. Recycled PP exhibited greater crystallization rate, higher crystallinity and equilibrium melting temperature than those measured for virgin PP. Elastic modulus and yield stress increase with the number of recycling steps, however, elongation at break and fracture toughness decrease [67].

Martins and De Paoli [68] investigated extrusion moulding for 13 processing cycles, using formulations containing virgin PP and 1:1 mixtures of virgin and post-consumer PP containing, or not, antioxidant. The main effect of reprocessing for PP was the decrease of molar mass due to degradation with chain scission, which was detected by the melt flow rate and the increase in crystallinity. Antioxidant consumption and material stability reduction with processing cycles were observed by the reduction of the thermo-oxidative stability detected by oxidative induction time measurements. The addition of antioxidants minimized degradation and maintained the stability of the material.

3.5.2. The Effect of Processing on Properties

Nasir et al. [69] simulated the thermo-oxidative degradation of PP during service life and recycling. Injection molded specimens were heat aged at 130°C for different times up to maximum of 300h to simulate the degradation of PP during the service life. These aged specimens were mixed with stabilizers in internal mixer and again heat aged up to 300h. A small increase in melt flow rate value was observed for aged PP but it showed large increase after recycling. Carbonyl index (CI) was increased in recycled sample with aging, whereas oxidation induction time (OIT) decreased. The stabilizers used during reprocessing were quite effective in controlling the thermo-oxidative degradation of the polymer during processing and aging.

The thermal properties are of importance for semicrystalline polymers such as PP for an understanding of the crystallinity and crystalline structure of the waste product. The crystallinity of PP depends on the molecular mass, the molecular mass distribution, the presence/absence of comonomers and additives/foreign particles, and on processing history.

Rabello and White observed that the melting enthalpy of PP increased on increasing the time of ultraviolet exposure up to a critical time but remained unchanged beyond this critical exposure time [70]. This was suggested to be due to a competition between two aspects of degradation, namely, reduction in molecular mass and formation of chemical impurities such as carbonyl and hydroperoxide groups. At the initial stages the reduction in molecular mass was said to be more dominant and enthalpy increased but with longer exposure, crystallization process was hampered by the presence of chemical impurities and therefore the enthalpy remained unchanged beyond the critical time.

PP post consumer waste (PP PCW) differs from reprocessed and artificially degraded PP in that PCW is exposed to a range of phenomena including thermal and mechanical loading during the manufacture, filling, transport, storage, use and disposal at the municipal recycling facilities (MRF). In addition, these materials suffer from contamination. DSC has been used by Majumdar et al. to study the heat flow during melting and crystallization of a range of PP PCW grades and blends [71]. They observed that the enthalpies of PP PCW are significantly lower than those for comparable virgin PP. This reduction was attributed to degradation of these materials during residence at the MRFs. Under the influence of UV radiation the PP chains undergo chain scission, which forms chain defects that reduces the number of perfect crystallisable PP chains in the materials and therefore the enthalpies of PP PCW were reduced compared with virgin PP. The PP PCW materials contain polymeric contaminants such as LDPE, LLDPE and HDPE even after vigorous sorting and cleaning process. The enthalpies of PP PCW, however, are not altered by low amounts (up to 10%) of polymeric contaminants. The enthalpies of the PCW blends could in general be predicted by a linear additive rule, which is of importance for recycling a variety of PP PCW products [71].

Techniques, for instance automatic sorting facilities using infra red technology, can be used to reduce the amount of impurities, but their use increases the cost significantly. Recycled PP has lower fracture properties compared to virgin material. Virgin PP homopolymer has a rather low modulus, a limited impact resistance and high notch sensitivity. To improve the mechanical and thermal properties, additives, such as mineral fillers and rubbers or elastomers, are blended with PP. Elastomer such as ethylene-propylene rubber (EPR) is used to enhance toughness and impact resistance, but has the side effect of also reducing the yield strength and Young's modulus. To complement as well as amplify the effect of the elastomer content, inorganic filler was added. Literature refers to two separate possibilities for where the inorganic fillers end up, the elastomer particles and rigid particles are scattered in PP matrix independently, or the elastomer particles attract and absorb the rigid particles [72].

Brachet et al. [73] studied the mechanical and thermal properties of recycled PP from post-consumer containers with addition of stabilizers, elastomer (ethylene-octene rubber, EOR) and calcium carbonate. The results showed limited changes with the addition of elastomer and $CaCO_3$ on the mechanical properties of recycled PP. The PE impurities and the presence of many different qualities of PP in the recycled material may have prevented any possible improvement in the mechanical properties by the addition of EOR and $CaCO_3$, improvements seen on virgin PP. They concluded that reusing and upgrading of recycled PP from post-consumer containers would first require a better sorting of the post-consumer waste. In addition the use of an adequate compatibilizer that would allow a uniform and homogeneous blending of the raw material mixture might enhance the mechanical properties [73].

3.5.3. Chemical Treatment and Pyrolysis

Chemical recycling, i.e., thermal and/or catalytic conversion of waste polymers into fuels or chemicals, has been recognised as a potential source of chemicals and energy that could significantly reduce the net cost of disposal. The most widely conventional chemicals methods for waste polymer treatment are pyrolysis and catalytic reforming. Pyrolysis, i.e., thermal cracking of the polymers to give low molecular weight materials, is attractive, but unfortunately gives a very broad range of products. It also requires high temperatures typically more than 500°C and even up to 900°C.

Catalytic degradation offers considerable advantages since suitable catalysts have the ability to control both the product yield and product distribution from polymer degradation as well as to reduce significantly the reaction temperature, potentially leading to a cheaper process with more valuable products. Catalytic pyrolysis has been carried out by considering a variety of catalysts with little emphasis on the reactor design, with only simple adiabatic batch and fixed-bed reactors being used. Due to the lack of intimate contact between polymer and catalyst over the whole reactor on the above systems, the option of using fluidized-bed reactor to expand the performance of fluidized catalytic cracking (FCC) catalysts in polymer degradation became attractive.

Lin and Yang [74] studied the PP pyrolysis using FCC commercial catalysts (FCC-s1) in a laboratory fluidised-bed reactor operating isothermally in the temperature range of 330-450°C at ambient pressure. The yield of gaseous and liquid hydrocarbon products at 390°C for spent FCC catalyst (87.8 wt%) gave much higher yield than silicate (17.1 wt%). Greater product selectivity was observed with FCC-s1 as a post-use catalyst with about 61 wt% olefins products in the C_3-C_7 range. Valuable hydrocarbons of olefins and iso-olefins were produced by low temperatures and short contact times [74].

Pyrolysis can tolerate mixtures of different plastic resins as well as high level of organic contamination which is also pyrolyzed. It has been reported that the addition of some PS to polyolefin feedstock (PE, PP) accelerates the pyrolysis and leads to decomposition of products with a shorter chain length, e.g. more volatile products. More oil was obtained than expected form addition of both individual polymers [75].

Pyrolysis of a waste plastic mixture (HDPE, LDPE, PP, PS =3:2:3:1) into a liquid product was also carried out in a semi-batch reactor at low and high temperatures [76, 77].

Siddiqui and Redhwi [78] studied the thermal and catalytic pyrolysis of LDPE, HDPE, PP, PET and PS individually and in mixed ratios with PS. Their results indicated that ratio 1:1 afforded best yields in the form pyrolytic oils. SIM distillation of hexane soluble portions showed that the low boiling fractions were not found and fractions were obtained only after 96°C boiling point. In most of the binary pyrolysis, light cycle oil fractions have low recovery than heavy cycle oil. Mixture of LDPE/HDPE/PP was pyrolyzed over various catalysts using a laboratory fluidized-bed reactor operating isothermally at ambient pressure [79]. It was observed that the yield of volatile hydrocarbons for zeolitic catalysts gave higher yield than non-zeolitic catalysts. It was also found that conversion of post-consumer LDPE/HDPE/PP polymer waste to volatile hydrocarbons was more than 80% of feed in 20 min in the presence of cracking catalysts with fluidized-bed reactor used in this study. Combustion destructs the solid plastics and generates a lot of heat energy. However, plastic combustion generates toxic species, such as polychlorinated dibenzo-dioxin/polychlorinated dibenzo-furan, soot, and polycyclic aromatic hydrocarbons. Han et al. [80] studied the combustion of PP in a semi-pilot scale fluidized bed combustor with porous alumina bed material. Their results showed that both excess air and temperature increase can suppress the emission of CO, while NO_x and SO_2 emission was independent of excess air and temperature.

B. Experimental Results from Various Recycling Systems

In the above review it was shown that, in general, a deterioration of physicomechanical properties of plastics recycled from WEEE can be recorded. This effect is a result of the sum of all types of ageing effects occurring during the product's life cycle plus the degradation in manufacturing processes and weathering during the product's use. In addition, the presence of impurities that are difficult to remove from a recycling stream, such as platings, labels and paintings, further contribute to the decrease of critical properties. More specifically, the above survey presents an analysis of the effect of recycling cycles on mechanical and thermal properties, as well as on the rheology and thermal response of recycled polymers. However, the overall estimation of recycling plastics from WEEE reveals a clear superiority of this approach, since the CO_2 emissions of recycled materials are smaller than those of the virgin at any rate of the process. The reason is that CO_2 emissions in the recycling process are smaller than those corresponding to the manufacturing of virgin polymer, starting from mining of raw materials to polymerization. Therefore, polymer recycling and further reuse of the recycled products for the design of new components is steadily increasing and it is widely accepted that direct use of post-consumer polymers is the most efficient way to treat plastic waste. However, the performance of products made of recycled polymers will differ from products made of virgin grades, because the initial processing step creates the above mentioned degradation of properties via mechanisms and reactions that are often detrimental to the properties of polymer. Attempts have been made to overcome this problem and improve final performance of the recycled plastics, either by the use of additives such as reinforcements or by adjusting an optimized consistency of the plastic waste stream. This latter procedure has the additional advantage of using. A mixture without separation, usually being a costly and time consuming stage. For instance, when recycling mineral oil bottles and containers, a mixture of polyethylene/polypropylene is obtained, that normally could be processed without separation, provided that enough compatibilization can be promoted by additives. This will enable us to use the as received stream and furthermore to obtained recycled products with improved properties due to synergistic effect of the blend components. By the following experimental work, mixtures of polymers with predominant concentration in WEEE were investigated and the effect of processing on their structure parameters and performance properties was explored.

1. Materials and Methods

1.1. Preparation of Blends

PC/ABS, HIPS/ABS and HDPE/PP blends were prepared using a co-rotating twin screw extruder with L/D-25 and 16mm diameter (Haake PTW 16).

It is critical to adjust an optimum extrusion temperature for processing of the polymer blends, in order to avoid either under-plasticization due to low temperature or thermal degradation because of over-exposure of the melt.

A circular die with an opening of 2mm diameter was mounted to the extruder. The obtained polymer blend, in the form of continuous strand was granulated using a Brabender pelletizer.

1.2. Injection Moulding

Injection moulding was performed with an ARBURG 221K ALLROUNDER machine, equipped with a screw of 25mm diameter and clapping unit force of 350kN. The mould was a two-cavity system capable of producing dump-bell specimens for subsequent measurement of tensile properties (ASTM D-638, specimens type IV).

1.3. Differential Scanning Calorimetry (DSC)

DSC measurements were run in a Mettler Toledo model DSC 1 differential scanning calorimeter with pure indium as a calibration standard. All runs were conducted under nitrogen flow of $20cm^3$/min to limit thermooxidative degradation.

1.4. Thermogravimetric Analysis (TGA)

Thermogravimetric analyses of samples were performed in Mettler Toledo thermogravimetric analyzer (model TGA-DTA). The tests were run with samples of 10 mg, at a heating rate of 10°C/min from 25°C to 700°C, in a nitrogen atmosphere.

1.5. Melt Flow Index (MFI)

The MFI measurements were carried out according to procedure A of the ASTM D 1238 specification, in a Kayeness Co. model 4004 capillary rheometer.

1.6. Mechanical Tests

Tensile testing was carried out according to ASTM D-638 specification. Five specimens from each composition (with dimensions according to type IV specimen of the above specification) were tested using an Instron (model 4466) tensometer, equipped with a load cell of maximum capacity of 10kN, operating at a grip separation speed of 50mm/min.

2. Results and Discussion

2.1. ABS/PC Blends

The terpolymer poly(acrylonitrile – butadiene – styrene) (ABS) was supplied by BASF, under the trade name Terluran® GP-35. The polycarbonate was Makrolon® 2805 by Bayer.

2.1.1. Differential Scanning Calorimetry (DSC)

The DSC analysis aimed to verify the miscibility of the ABS/PC blends. In polymeric materials the T_g is particularly related to the mobility of the polymeric chains and determines the transition between the glassy and the rubbery polymeric state. If the system is immiscible, T_g values of individual polymers will nor change, otherwise if the system is completely miscible, it will display only one T_g. Figure 1 shows the DSC thermographs of ABS, PC and their examined blends.

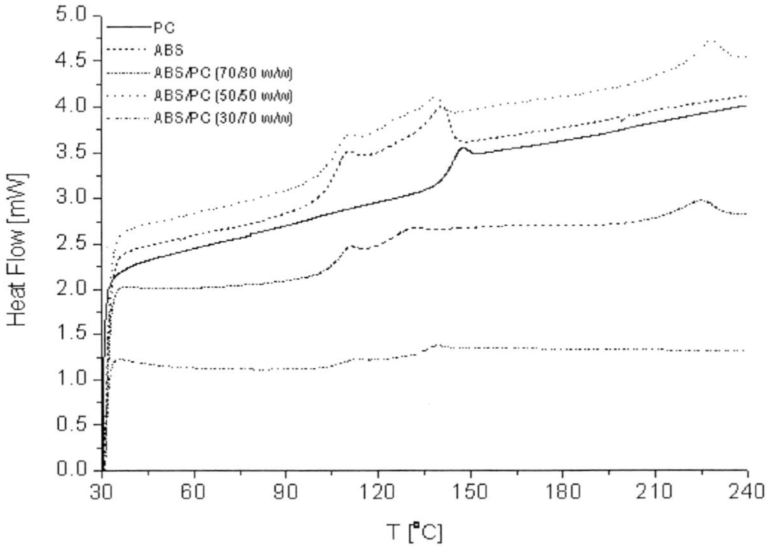

Figure 1. DSC thermographs of ABS/PC blends.

The glass transition temperatures of ABS/PC blends, as derived from the DCS runs, are listed in Table 1. At blend composition ABS/PC 70/30 w/w, a single transition point was recorded corresponding to the polymer of higher concentration in the blend (ABS). For sample compositions ABS/PC 50/50 and 30/70 w/w, the results of this Table show two glass transitions, each one corresponding to the respective pure component. However, a shift of both T_g can be observed in the range of a few degrees, and more specifically the first transition takes place at a temperature slightly higher than that corresponding to pure ABS and shifts further to higher temperatures as the PC content of the blend increases.

The second transition of the blend appears at a temperature slightly lower that the T_g of pure PC that further decreases as the ABS content in the blend increases. These interactions can be clearly seen in Figure 1, where the results obtained from the thermographs are plotted versus the temperature range of scan. The dependence of T_g values with the composition indicates certain interactions between the components.

Table 1. DSC analysis data of ABS/PC blends

ABS/PC (w/w)	T_{g1} (°C)	T_{g2} (°C)
100/0	105.03 ±0.09	-
70/30	105.31 ±0.26	-
50/50	106.35 ±0.21	124.35 ±0.56
30/70	108.31 ±0.10	134.19 ±0.82
0/100	-	142.77 ±0.04

In fact, if the system was immiscible, T_g values of individual polymers would remain unaffected. On the other hand, a completely miscible mixture would display only one T_g. It also should be noted that the shift of T_g is much more enhanced for PC, where a decrease from 142.77 to 124.35°C was recorded. This can be interpreted assuming a plasticizing effect of ABS on PC, which is likely to happen due to the polybutadiene phase present in ABS.

This indicates the partial miscibility of the above blends, since the decrease in the T_g of the PC indicates that the two materials tend to form a miscible mixture. The above results are in agreement with those reported by Balart et al. [43], who also run miscibility studies for ABS/PC blends.

This partial miscibility was attributed to a low interaction between the phases, promoting a decrease in mechanical performance. This trend, indicates the existence of a wide range of miscibility of these two components, depending on their proportion in the composition

Balart et al. [43], used the Fox equation for calculation of the weight and volume fractions of the miscible segments (PC and SAN), assuming that butadiene phase does not take part in miscibility interactions, because of its net structure.

They concluded that PC is able to incorporate higher SAN amounts, due to the formation of low molecular weight species on SAN, as derived from degradation during pre-processing and service life.

2.1.2. Thermogravimetric Analysis (TGA)

The thermal stability of ABS/PC blends was assessed be means of TGA (Figure 2). The onset temperature of thermal degradation (T_{onset}) and the temperature corresponding to maximum reaction rate (T_{max}) were calculated (Table 2).

Determination of thermal properties of ABS-PC blends obtained from WEEE using TGA by Balart et al. showed the validity of the autocatalytic model [43].

They observed that in ABS-based mixtures the activation energies of thermal degradation were independent of heating rates which suggested that the most important process in degradation of these materials corresponds to ABS, in particular thermo-oxidative processes in the elastomeric phase of the blend.

Mixtures with high PC content showed a clear increase in apparent activation energy with heating rate, suggesting that the thermal degradation mechanism of these samples is composed of several complex processes, each predominant during different stages of the overall process. They also mentioned that the addition of a high amount of ABS permits the creation of a network formed by elastomeric domains which lead to a general increase in heterogeneity of the mixture.

Table 2. TGA data of ABS/PC blends

ABS/PC	T_{onset} (°C)		T_{max} (°C)	
(w/w)	Experimental values	Rule of mixt.	Experimental values	Rule of mixt.
100/0	402.23 ±0.11	-	422.64 ±0.50	-
70/30	408.32 ±0.81	432.70	429.80 ±0.03	453.17
50/50	410.98 ±1.34	453.02	430.83 ±1.22	473.53
30/70	416.05 ±1.12	473.33	439.97 ±1.85	493.89
0/100	503.80 ±0.79	-	524.42 ±0.10	-

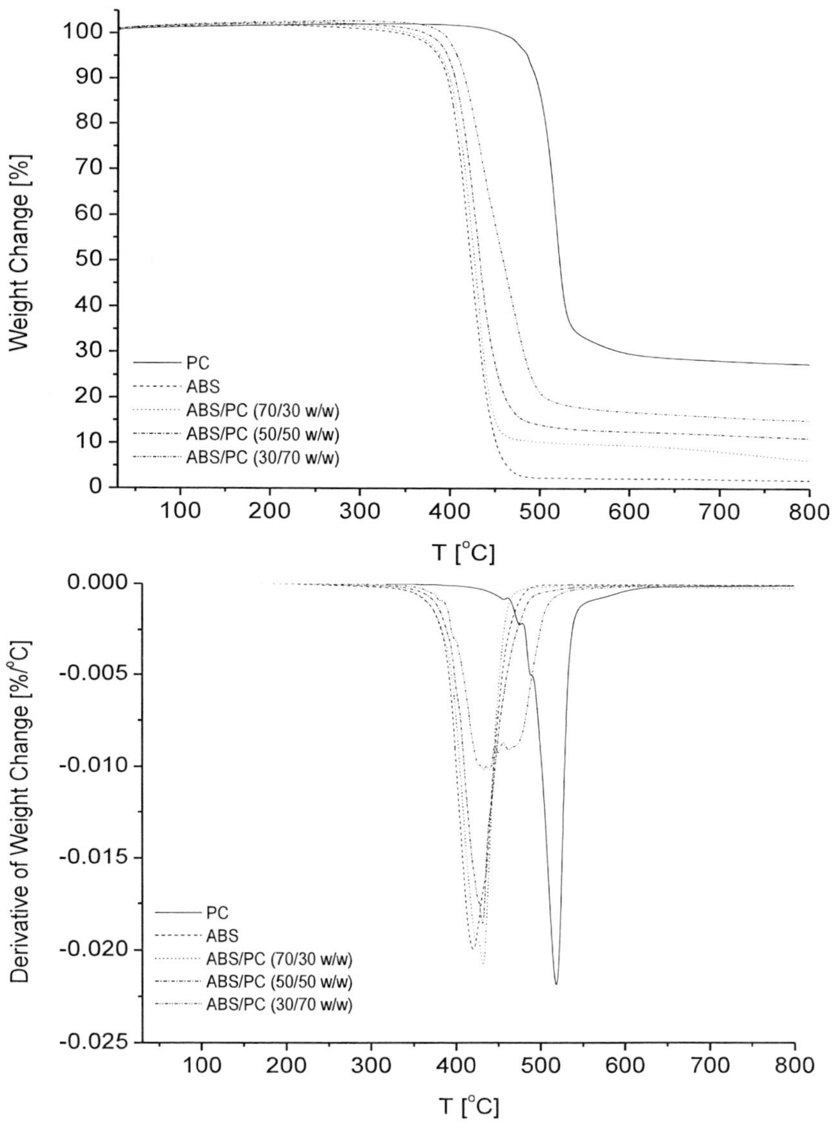

Figure 2. Thermogravimetric analysis of ABS/PC blends.

2.1.3. Melt Flow Index (MFI)

For assessment of the data obtained from MFI measurements, it should be kept in mind that PC is homogeneous, singe-phase polymer, while ABS is a heterogeneous terpolymer consisting of a rubbery polybutadiene phase dispersed in a continuous plastic phase of SAN.

The incorporation of PC polymer into ABS results in an increase of the ratio of plastic/rubbery phase, which modifies the interactions between these phases and further changes the structural morphology of the system resulting in different rheological properties. From the results of Table 3 it can be observed that the MFI values of PC and ABS differ significantly, and therefore, most of the mixtures show rheological behaviour similar to that of PC. Very interestingly, the blend containing 50% ABS shows a significant increase of MFI, which suggests that this blend can be easily processed in comparison with PC alone. This can probably be interpreted by the fact that ABS plays the role of plasticizer at that concentration, due to the elastomeric phase interactions. Similar results have been reported by Khan et al. [81], who studied the rheological properties of ABS/PC blends and observed that the shear thinning behaviour of ABS improves the processability of PC. This effect provides a significant power saving during melt processing of these blends for mixing or moulding purposes. Moreover, plasticization of PC by the addition of ABS is of great importance to the recycling procedures of plastics from WEEE. In fact, recycling of these plastics requires a minimal exposure to high temperatures and shear in order to avoid extensive decomposition of the polymers due to thermo-oxidative and thermo-mechanical degradation.

Table 3. MFI measurements of ABS/PC blends

ABS/PC (w/w)	MFI (g/10min)
100/0	2.00 ±0.16
70/30	12.91 ±0.42
50/50	16.30 ±0.39
30/70	16.22 ±0.56
0/100	5.76 ±0.55

2.1.4. Tensile Tests

PC exhibits the highest yield strength and ductile behavior. In contrast, ABS shows a brittle character and has the lowest yield strength. From Table 4 it can be observed that the incorporation of ABS into PC is accompanied by a decrease in both tensile strength and ductility of the blend in comparison with pure PC.

In fact, the tensile strength decreases as the concentration of ABS increases in the polymer blend, as expected by the law of mixtures. A decrease of mechanical properties of ABS/PC blends deriving from recycled WEEE was also reported by Balart et al. [82]. The above authors interpreted this behaviour by the thermo-oxidative degradation process, taking place in the sensitive elastomeric phase (polybutadiene rubber dispersed in SAN). The partial miscibility of the ABS/PC system is an additional reason for this effect. In fact, it was found that SAN phase is not completely miscible with a PC rich phases and the lack of adherence between these two conjugated phases can influence negatively the mechanical performance. In agreement with our results, they indicated as optimum blend composition in terms of mechanical performance, the range of PC concentration from 10 to 20%.

Table 4. Tensile test results for ABS/PC blends (in parenthesis values calculated by the "rule of mixtures")

ABS/PC (w/w)	Tensile strength (MPa)	Modulus of elasticity (MPa)	Strain at break (%)
100/0	46.77 ±0.22	2105.42 ±70.44	13.21 ±3.15
70/30	51.96 ±0.62 (52.16)	2181.15 ±103.40 (2053.79)	13.04 ±1.22 (16.02)
50/50	57.00 ±0.80 (55.75)	2358.99 ±27.36 (2019.37)	17.04 ±3.58 (17.89)
30/70	58.87 ±0.43 (59.34)	2113.00 ±69.16 (1984.95)	64.22 ±13.30 (19.76)
0/100	64.72 ±0.52	1933.32 ±87.28	22.57 ±3.04

The rule of mixture for tensile yield strength of ABS/PC blends was confirmed by Khan et al. [81], but it was also found that the changes of elongation at break and impact strength do not follow this rule. The above researchers concluded that the blend morphology plays a significant role on the determination of mechanical characteristics of the samples.

In another work, Wong [83], reports that the value of tensile modulus of the system ABS/PC, as well as its increasing or decreasing tendency, depend on the viscosity of ABS. This statement could be correlated to the above mentioned hypothesis that ABS produces a plasticizing effect, when incorporated to PC.

2.2. ABS/HIPS

The following commercial grades of engineering thermoplastics used in electrical and electronic application were studied:

An easy flowing ABS grade, suitable for injection moulding, with trade name Terluran GP-22 (BASF, Germany), exhibiting high resistance to impact and heat deflection. Styron A-TECH 1200 (Dow Hellas, Greece), a HIPS grade with good combination of toughness and stiffness, good processability and specially designed for packaging applications.

2.2.1. Differential Scanning Calorimetry (DSC)

The results obtained from the DSC thermographs of ABS/HIPS blends (Figure 3) are shown in Table 5 and display a single T_g peak. According to the above discussion this suggests that all compositions of this mixture are fully miscible.

The results of Table 5 show that T_g of ABS decreased with increasing proportion of HIPS in the prepared blend.

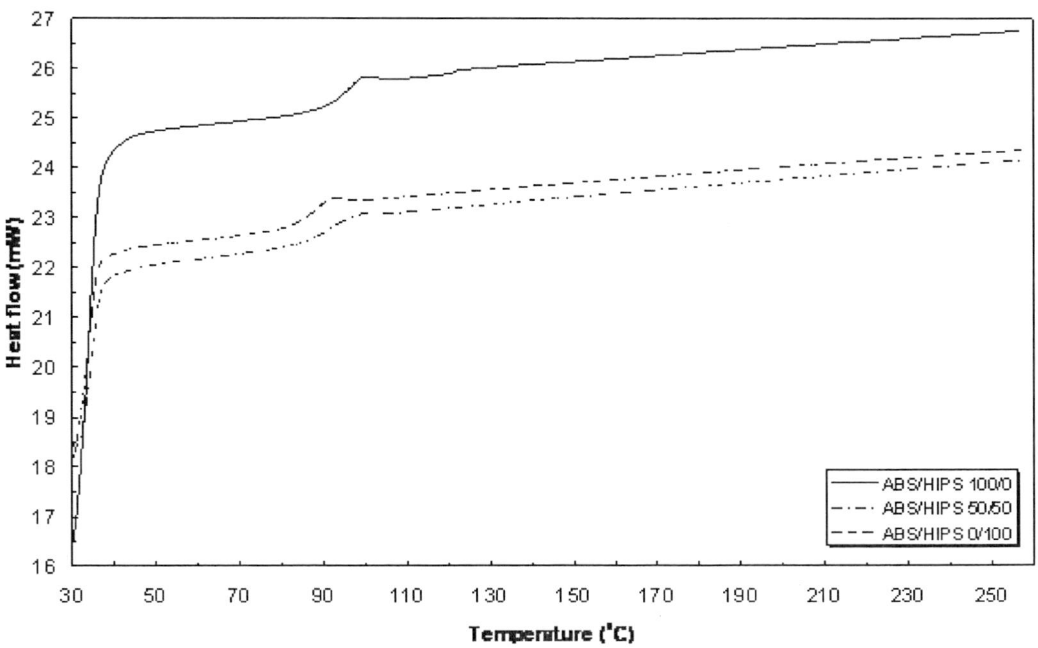

Figure 3. Comparative thermographs of ABS/HIPS blends (2nd heating cycle).

Table 5. DSC analysis data of ABS/HIPS blends

ABS/HIPS (w/w)	T_g (°C)
0/100	85.83
20/80	87.00
50/50	91.10
80/20	93.66
100/0	95.12

2.2.2. Melt Flow Index (MFI)

From Table 6 it can be seen that the MFI values increase by increasing HIPS concentration in the polymer blend, as it was expected by the application of the "rule of mixtures". In that case, we cannot see any plasticizing effect because both components of the mixture, i.e. ABS and HIPS, contain a rubbery phase, probably acting already as a plasticizer.

The investigation of rheological characteristics along with some structure parameters and final properties of the ABS/HIPS mixture could provide useful information for their recycling process.

In addition, ABS/HIPS mixtures could be suitable for use as additives in PC recycling, instead of ABS alone, contributing to the low cost of final products. This is reasonable taking into account that the dispersed rubber phase in HIPS could act in a similar mode as the butadiene phase present in ABS, i.e. as shock absorber that greatly contributes to impact resistance of the recycled mixture of plastics.

Table 6. MFI measurements of ABS/HIPS blends

ABS/HIPS (w/w)	MFI (g/10min)	
	Experimental values	"rule of mixtures"
0/100	5.669 ±0.355	-
20/80	6.029 ±0.192	5.258
50/50	4.225 ±0.231	3.962
80/20	2.321 ±0.090	2.687
100/0	1.836 ±0.127	-

Table 7. Tensile test results for ABS/HIPS blends

ABS/HIPS (w/w)	Tensile strength (MPa)	Modulus of elasticity (MPa)	Strain at break (%)
0/100	32.40 ±1.34	1322.60 ±41.65	40.47 ±3.28
20/80	34.89 ±0.65	1409.75 ±14.21	39.71 ±4.52
50/50	40.87 ±0.51	1975.57 ±56.82	22.60 ±5.58
80/20	47.42 ±0.39	2045.14 ±61.42	20.77 ±5.58
100/0	50.65 ±4.74	1918.57 ±81.85	4.41 ±2.00

2.2.3. Tensile Tests

ABS was found to have tensile strength greater than that of HIPS (Table 7). The tensile strength results for the blends ABS/HIPS follow the "rule of mixtures" showing a decreasing trend with the increase of HIPS content. However, a synergistic effect can also be observed in the modulus of elasticity of ABS/HIPS blends, which is more significant at the composition of ABS/HIPS: 80/20.

From the above table, it is obvious that the addition of HIPS in ABS increases the ductility of the obtained material, in comparison with pure ABS. These results are in agreement with the research work of Brennan et al. [20] on the effect of mixing ABS with HIPS on their mechanical properties. They came to the conclusion that blending ABS and HIPS caused no more deterioration in properties than that occurred as a result of the recycling process, and the presence of small content of one material in the other restored significant amounts of ductility.

2.3. HDPE/PP

HDPE (Borealis) and PP (MIDILENA III-F401, Rompetrol Petrochemicals S.R.L.) blends were prepared with a co-rotating twin screw extruder, at 30 rpm. The following blends with 10, 20 and 25% (w/w) HDPE in PP were prepared.

2.3.1. Differential Scanning Calorimetry (DSC)

The presence of two peaks in the DSC thermographs reveals uncompatibility between the HDPE and PP phase of the polyolefin blends (Figure 4). From the DSC results (Table 8) a decrease in the melting temperature of PE phase (T_{m1}) of the prepared blends can be observed, in comparison with the pure HDPE. The melting temperature (T_{m2}) and the enthalpy (ΔH_2) of PP phase remain almost constant.

Table 8. DSC analysis data of HDPE/PP blends

HDPE/PP (w/w)	T_{m1} (°C)	ΔH_1 (cal/g)	T_{m2} (°C)	ΔH_2 (cal/g)
0/100			166.42 ±0.58	13.58 ±0.44
10/90	133.24 ±0.61	2.34 ±0.46	164.40 ±0.64	12.33 ±0.37
15/85	133.65 ±0.80	4.86 ±0.49	163.33 ±1.07	11.05 ±0.60
20/80	134.56 ±0.40	6.60 ±1.07	164.45 ±1.25	10.24 ±0.55
25/75	135.00 ±0.59	7.14 ±0.34	164.85 ±0.50	10.45 ±1.33
100/0	136.67 ±0.70	36.05 ±2.81	-	-

2.3.2. Thermogravimetric Analysis (TGA)

TGA showed that the incorporation of HDPE in PP improves the thermal stability of the prepared blends with respect to pure PP, in terms of both temperatures onset (T_{onset}) and maximum rate of thermal degradation (T_{max}), according to the "rule of mixtures" (Table 9).

2.3.3. Melt Flow Index (MFI)

From the results of Table 10 it is observed that the addition of HDPE decreases the MFI values of the prepared blend, in comparison with pure PP, according to the "rule of mixtures".

2.3.4. Tensile Tests

Tensile test measurements in HDPE/PP blends (Table 11) showed a tendency for increase in modulus of elasticity whereas no obvious effect in tensile strength was recorded. A slight increase of the strain at break was observed which reveals that better drawing behaviour of the prepared HDPE/PP blends.

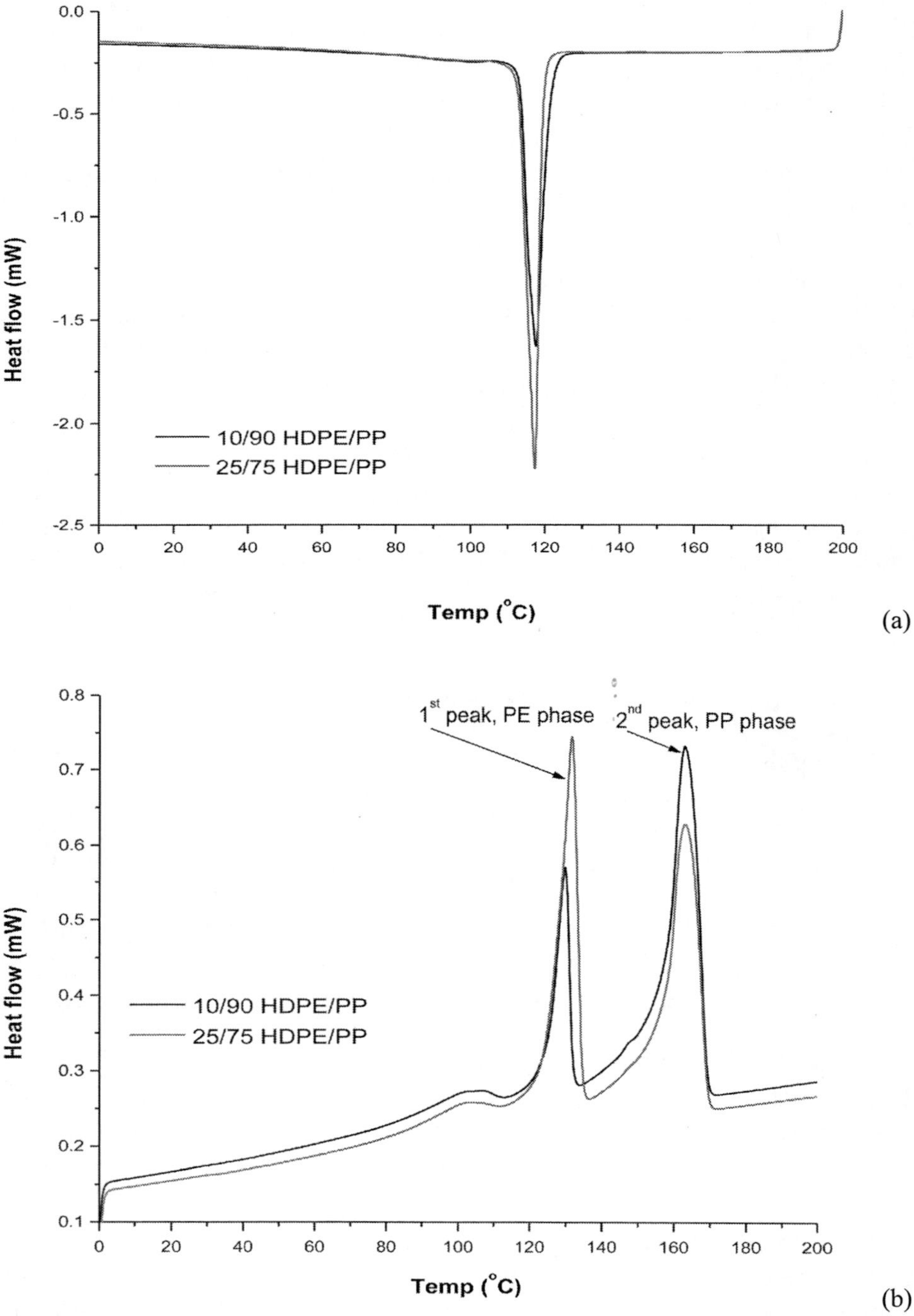

Figure 4. DSC thermographs of HDPE/PP blends during: (a) 2nd heating and (b) cooling cycle.

Table 9. TGA data of HDPE/PP blends

HDPE/PP (w/w)	T_{onset} (°C)	T_{max} (°C)	Residue (%)
0/100	392.0	433.9	4.47
10/90	401.3	441.3	3.55
15/85	422.7	451.0	2.11
20/80	432.4	456.4	4.33
25/75	413.6	450.7	3.77
100/0	456.3	476.4	3.47

Table 10. MFI data of HDPE/PP blends

HDPE/PP (w/w)	MFI (g/10min)	
	Experimental value	Rule of mixture
0/100	2.69 ±0.0043	-
10/90	2.43 ±0.0045	2.52
15/85	2.39 ±0.0056	2.43
20/80	2.38 ±0.0061	2.35
25/75	2.15 ±0.0062	2.26
100/0	0.98 ±0.0015	-

Table 11. Tensile test results of HDPE/PP blends

HDPE/PP (w/w)	Tensile strength (MPa)	Modulus of elasticity (MPa)	Strain at break (%)
0/100	38.80 ±0.52	1165.88 ±57.96	19.61 ±2.93
10/90	39.40 ±0.35	1207.83 ±42.25	17.81 ±1.29
15/85	39.60 ±0.36	1205.26 ±28.55	19.07 ±0.65
20/80	39.09 ±0.28	1223.26 ±39.75	20.25 ±4.59
25/75	38.80 ±0.46	1199.21 ±21.41	20.48 ±0.91
100/0	27.48 ±0.40	764.64 ±34.56	31.56 ±3.20

CONCLUSION

The blends of engineering plastics studied in this work show interesting structure/property interrelations and more specifically:

PC/ABS mixtures display partial miscibility, as shown by the results of differential scanning calorimetry. Regarding their microstructure, the above mixtures contain PC rich zones including some ABS chains as well as ABS rich zones where some PC chains are also present in the SAN matrix. Very interestingly, ABS seems to act as plasticizer when incorporated into PC into concentrations of about 50%, probably due to the dispersion of the polybutadiene phase within the thermoplastic matrix of the blend. This effect was also clear during the determination of rheological characteristics of the PC/ABS blend via MFI measurements. Here again, the sample with composition 70/30 shows increased flow rate, which is consistent with the hypothesis of plasticization and suggests easier processing in the industrial scale (e.g. during extrusion or injection moulding). On the other hand, the mixtures of HIPS with ABS seem to produce fully miscible blends and, furthermore, the determined changes of blend properties are mostly in good agreement with values calculated from the law of mixtures.

A decrease of mechanical properties, in terms of tensile strength and elongation, was recorded for both types of mixtures i.e. ABS/PC and ABS/HIPS, with respect to the properties of pure polymers. However, the modulus of elasticity shows some improvement in both cases, due to the increase of stiffness of the corresponding blends. The results obtained from this study allow us to assume that mixtures of ABS with PC or HIPS can be directly prepared using the mixed engineering plastics stream coming from the recycling of WEEE and these materials will show easy processing, acceptable mechanical properties and low cost, providing at the same time an interesting approach to management of solid plastic waste deriving from electrical and electronic devices.

As far as the polyolefin blends is concerned, the incorporation of HDPE in PP improves its thermal stability. The thermal properties of PP phase remain unchanged. The tensile strength and modulus of elasticity are not significantly affected, whereas the strain at break increases which is an indication of better deformation ability of these blends.

The experimental design in the above presented work, might be a pilot methodology for the exploration of the performance of various polymeric blends deriving from the as received plastic waste streams. Mixtures containing more than two components would also be an interesting field of research, since these polyblends might exhibit unexpectedly improved properties, even beyond those of their individual components. On the other hand, this study is probably of value, with respect to the effect and the changes taking place in the structure characteristics of the investigated mixtures, which is the background for an interpretation of the overall behavior of final products and, moreover, it can elucidate our knowledge about the structure-property interrelations established in those systems. Finally, the above studied procedures of preparing recycled polyblends seems to be significant for the industrial practice, as it can lead to final products that are suitable for various applications and display acceptable properties with minimal production cost.

REFERENCES

[1] Hopewell, J.; Dvorak, R.; *Kosior, E. Philosophical transactions of the Royal Society B* 2009, 364, 2115-2126.

[2] Al-Salem, S.M.; Lettieri, P.; Baeyens, *J. Progr. Energy Comb. Sci.* 2010, 36, 103-129.

[3] Al-Salem, S.M.; Lettieri, P.; Baeyens, *J. Waste Manag.* 2009, 29, 2625-2643.

[4] Fisher, M.M.; Mark, F.E.; Kingsbury, T.; Vehlow, J.; Yamawaki, *T. IEEE,* 2005, 83-92.

[5] Chancerel, P.; Rotter, S. *Waste Managem.* 2009, 29, 2336-2352.

[6] Dalrymple, I.; Wright, N.; Kellner, R.; Bains, N.; Geraghty, K.; Goosey, M.; Lightfoot, L. *Circuit World* 2007, 33, 52-58.

[7] Taurino, R.; Pozzi, P.; Zanasi, T. *Waste Managem.* 2010, 30, 2601-2607.

[8] Schlummer, M.; Mäurer, A.; Leitner, T.; Spruzina, W. *Waste Manag. Research* 2006, 24, 573-583.

[9] Dimitrakakis, E.; Janz, A.; Bilitewski, B.; Gidarakos, E. J. *Hazardous Mater.* 2009, 161, 913-919.
[10] Dawson, R.B.; Landry, S.D. *IEEE* 2005, 77-82.
[11] Schlummer, M.; Mäurer, A. *J. Appl. Polym. Sci.* 2006, 102, 1262-1273.
[12] Tange, L.; Drohmann, D. *Polym. Degrad. Stab.* 2005, 88, 35-40.
[13] Bientinesi, M.; Petarca, l. *Waste Manag.* 2009, 29, 1095-1102.
[14] Luda, M.P.; Euringer, N.; Moratti, U.; Zanetti, M. *Waste Manag.* 2005, 25, 203-208.
[15] Marco, I. de; Caballero, B.M.; Chomón, M.J.; Laresgoiti, M.F.; Torres, A.; Fernández, G.; Arnaiz, S. *J. Analyt. Appli. Pyrol.* 2008, 82, 179-183.
[16] Yamawaki, T. *Fire Mater.* 2003, 27, 315-319.
[17] Cândido, L.; Kindlein, W.; Demori, R.; Carli, L.; Mauler, R.; Oliveira, R. *J. Clean. Product.* 2011, 19, 1438-1445.
[18] Iuga, A.; Samuila, A.; Neamtu, V.; Morar, R.; Beleca, R.; Das, S.; Dascalescu, L. *IEEE Transactions of Industry Applications* 2011, 47, 322-330.
[19] Yagi, S.; Furuhashi, M.; Inagaki, Y. *IEEE* 2005, 139-140.
[20] Brennan, I.B.; Isaac, D.H.; Arnold, J.C. *J. Appl. Polym. Sci.* 2002, 86, 572-578.
[21] Park, C.-H.; Leon, H.-S.; Yu, H.-S.; Han, O.-H.; Park, J.-K. Environm. *Sci. Technol.* 2008, 42, 249-255.
[22] Achilias, D.S.; Antonakou, E.V.; Koutsikosta, E.; Lappas, A.A. *J. Appl. Polym. Sci.* 2009, 114, 212-221.
[23] Garcia, D.; Balart, R.; Sánchez, L.; Lopez, *J. Polym. Engin. Sci.* 2007, 47, 789-796.
[24] Pérez, J.M.; Vilas, J.L.; Laza, J.M.; Arnáiz, S.; Mijangos, F.; Bilbao, E.; Rodríguez, M.; León, L.M. *J. Mater. Process. Technol.* 2010, 210, 727-733.
[25] Boldizar, A.; Möller, K. *Polym. Degrad. Stab.* 2003, 81, 359-366.
[26] Bai, X.; Isaac, D.H.; Smith, *K. Polym. Engineer. Sci.* 2007, 47, 120-130.
[27] Arostegui, A.; Sarrionandia, M.; Aurrekoetxea, J.; Urrutibeascoa, *I. Polym. Degrad. Stab.* 2006, 91, 2768-2774.
[28] Chen, S.-C.; Liao, W.-H.; Hsich, M.-W.; Chen, R.-D.; Lin, S.-H. Polym.-Plast. *Technol. Engin.* 2011, 50, 306-311.
[29] Karahaliou, E.-K.; Tarantili, P.A. *Polym. Engin. Sci.* 2009, 49, 2269-2275.
[30] Statler, D.; Stajduhar, E.; Gupta, R. *J. Fire Sci.* 2008, 26, 331-350.
[31] Chiu, S.-J.; Chen, S.-H.; Tsai, C.-T. *Waste Manag.* 2006, 26, 252-259.
[32] Watanabe, M.; Matsuo, Y.; Matsushita, T.; Inomata, H.; Miyake, T.; Hironaka, K. *Polym. Degrad. Stab.* 2009, 94, 2157-2162.
[33] Liu, F.; Li, Z.; Yu, S.; Cui, X.; Ge, X. *J. Hazardous Mater.* 2010, 174, 872-875.
[34] Endo, T.; Kakimoto, K.; Ochiai, B.; Nagai, D. *Macromolecules* 2005, 38, 8177-8182.
[35] Feller, J.F.; Bourmaud, A. *Polym. Degrad. Stab.* 2003, 82, 99-104.
[36] Pérez, J.M.; Vilas, J.L.; Laza, J.M.; Arnáiz, S.; Mijangos, F.; Bilbao, E.; León, L.M. *J. Polym. Environ.* 2010, 18, 71-78.
[37] Zenkiewicz, M.; Rytlewski, P.; Moraczewski, K.; Stepczyńska, M.; Karasiewicz, T.; Richert, J.; Ostrowicki, W.; Archiv. *Mater. Sci. Engin.* 2009, 37, 94-101.
[38] Drozdov, A.D.; Al-Mulla, A.; Gupta, R.K. *Mechan. Research Commun.* 2003, 30, 595-614.
[39] Sun, S.; He, Y.; Wang, X.; Wu, D. *J. Appl. Polym. Sci.* 2010, 116, 2451-2464.
[40] Nishihara, H.; Suda, Y.; Sakuma, T. *J. Fire Sci.* 2003, 21, 451-464.
[41] Liang, R.; Gupta, R.K. *ANTEC* (2002).

[42] Elmaghor, F.; Zhang, L.; Fan, R.; Li, H. *Polymer* 2004, 45, 6719-6724.

[43] Balart, R.; López, J.; Garcia, D.; Salvador, M.D. *Europ. Polym. J.* 2010, 41, 2150-2160.

[44] Eguiazábal, J.I.; Nazábal, *J. Polym Engin. Sci.* 1990, 30, 527-531.

[45] Tarantili, P.A.; Mitsakaki, A.N.; Petoussi, M.A. *Polym. Degrad. Stab.* 2010, 95, 405-410.

[46] Jin, D.W.; Shon, K.H.; Jeong, H.M.; Kim B.K. *J. Appl. Polym. Sci.* 1997, 69, 533-542.

[47] Khan, M.M.K.; Hilado, C.J.; Agarwal, S.; Gupta, R.K. *J. Polym. Environ.* 2007, 15, 188-194.

[48] Kalfoglou, N.K.; Chaffey, C.E. *Polym. Eng. Sci.* 1979, 19, 552-557.

[49] Soriano, F.; Morales, G.; Díaz de León, R. *Polym. Eng. Sci.* 2006, 42, 1698-1705.

[50] Vilaplana, F.; Ribes-Greus, A.; Karlsson, S. *Polym. Degrad. Stab.* 2006, 91, 2163-2170.

[51] Vilaplana, F.; Karlsson, S.; Ribes-Greus, A. *Europ. Polym. J.* 2007, 43, 4371-4381.

[52] Israeli, Y.; Lacoste, J.; Lemaire, J.; Singh, R.P.; Sivaram, S. *J. Polym. Sci.: Part A: Polym. Chem.* 1994, 32, 485-493.

[53] Parres, F.; Balart, R.; Crespo, J.E.; López, J. *J. Appl. Polym. Sci.* 2007, 106, 1903-1908.

[54] Dillon, P.S.; Aqua, E.N. Technical Report 20, *Chelsea Center for Recycling and Economic Development*, Univ. of Massachusetts, 2000.

[55] Xu, G.; Qiao, J.; Kuswanti, Ch.; Koelling, K.; Stuart, L.A.; Lilly, B. *J. Appl. Polym. Sci.* 2002, 84, 1-8.

[56] Tall, S.; Karlsson, S.; Albertsson, A.-C. *Polym. Polym. Compos. 1998*, 6, 261-267.

[57] Tang, H.Y.; Chaffey, C.E. *Plast. Rubber Process.*1989, 12, 53-61.

[58] Jamil, F.A.; Shubber, S.D.A. *Plast. Rubber Process. Appl.* 1988, 9, 187-190.

[59] Liu, X.; Bertilsson, H. *J. Appl Polym Sci.* 1999*,* 74, 510-515.

[60] Campomanes, R.M.; Manrich, S. *J. Appl. Polym. Sci.* 2003, 87, 747-751.

[61] Viplaplana, F.; Ribes-Greus, A.; Karlsson, S. *Anal. Chim. Acta* 2007, 604, 18-28.

[62] Mitan, N.M.M.; Bhaskar, T.; Hall, W.J.; Muto, A.; Williams P.T.; Sakata, Y. *Chemosphere,* 2008, 72, 1073-1079.

[63] Luzuriaga, S.; Kovářová, J.; Fortelný, I. *Polym. Degrad. Stab.* 2006, 91, 1226-1232.

[64] Luzuriaga, S.; Kovářová, J.; Fortelný, I. *Polym. Degrad. Stab.* 2011, 96, 751-755.

[65] Canevarolo, S.V.; Polym. *Degrad. Stab* 2000, 70, 71-76.

[66] Pfaendner, R.; Herbst, H.; Hoffmann, K.; Sitek, F. *Angew Makromol. Chem.* 1995, 232, 193-227.

[67] Aurrekoetxea, J.; Sarrionandia, M.A.; Urrutibeascoa, I; Maspoch, M.L.I. *J. Mater. Sci.* 2001, 36, 2607-2613.

[68] Martins, M.H.; De Paoli M.-A. *Polym. Degrad. Stab.* 2002, 78, 491-495.

[69] Nasir, A.; Yasin, T.; Islam, A. *J. Appl. Polym. Sci.* 2011, 119, 3315-3320.

[70] Rabello, M.S.; White, J.R. *Polymer* 1997, 38, 6389-6387.

[71] Majumdar, J.; Cser, F.; Jollands, M.C.; Shanks, R.A. *J. Therm. Anal. Calor.* 2004, 78, 849-863.

[72] Premphet, K.; Horanont, P. Polym. *Plast. Technol. Engin.* 2001, 40, 235-247.

[73] Brachet, P.; Høydal, L.T.; Hinrichsen, E.L.; Melum, F. *Waste Manag.* 2008, 28, 2456-2464.

[74] Lin, Y-H.; Yang, M.-H. *Polym. Degrad. Stab.* 2007, 92, 813-821.

[75] McNeill, I.C. Thermal Behavior of Mixtures of Two Plastics, Proc. 1st *Int. Symp. On Feedstock Recycling of Plastics,* Sendai Japan, Oct. 31-Nov. 3, 1999.

[76] Lee, K.H.; Shin, D.H. *Waste Manag.* 2006, 27, 168-176.

[77] Bhaskar, T.; Hall, W.J. Mitan N.M.M.; Muto, A. Williams, P.T.; Sakata, Y. *Polym. Degrad. Stab.* 2007, 92, 211-221.
[78] Siddiqui, M.N.; Redhwi, H.H. *Fuel Proces. Technol.* 2009, 90, 545-552.
[79] Wei, T.-T.; Wu, K.-J.; Lee, S.-L.; Lin, Y.-H. *Resourc. Conserv. Rec.* 2010, 54, 952-961.
[80] Han, J.; Shimizu, T.; Mataru, M.; Kim, H. *Energ. Sources, Part A* 2010, 32, 1121-1129.
[81] Khan, M.M.K.; Liang, R.F.; Gupta, R.K.; Agarwal, S. *Korea-Australia Rheolog. J.* 2005, 17, 1-7.
[82] Balart, R. ; Sánchez, L.; López, J.; Jiménez, A. *Polym. Degrad. Stab.* 2006, 91, 527-534.
[83] Wong, A. C.-Y. Polym-Plast. *Technol. Engin.* 2003, 42, 171-180.

In: E-Waste: Management, Types and Challenges
Editors: Yuan Chun Li and Banci Lian Wang

ISBN 978-1-61942-217-9

Chapter 2

CHALLENGES AND OPPORTUNITIES OF E-WASTE MANAGEMENT

A. Bazargan, K. F. Lam and G. McKay*
Department of Chemical and Biomolecular Engineering,
Hong Kong University of Science and Technology, Clearwater Bay, Hong Kong

ABSTRACT

In this review, the global production of e-wastes is described based on recent surveys and statistics. The variety and projections in e-waste composition is also identified due to the technological improvement in the electronic industry, which introduces difficulty for the development of a universal, sustainable recovery process. The effects of various organic and metal components of e-wastes on human health and the environment are summarized based on the literature findings. This review includes the challenges related to the rapidly increasing e-waste generation and describes the dilema that the developing countries are facing and the opportunities brought along with e-wastes to these countries. The chapter also includes a proposed process for the treatment and recocery of Printed Circuit Boards (PCBs).

With advancements in the electronic world almost occurring on a day-to-day basis and increased availability of products to the public, it is not surprising to see a staggering increase in the generation of electronic wastes over the past several decades. According to a United Nations Environment Program (UNEP) alert bulletin in 2006, in the manufacturing industry of industrialized countries, the production of electrical and electronic devices is the fastest-growing sector of industrial development. Thus the handling and treatment of electronic waste, or so-called E-waste, is a topic of concern in the future. Globally, there is no doubt that the amount of E-waste being produced and in turn disposed is sharply increasing. Although it is hard to give an accurate estimate of the global E-waste production due to faulty and sometimes non-existent data, the UN estimate of the global E-Waste production in the same alert was 20-50 million tons per year. In 1998, the E-waste production in the United States was estimated at 6 million tons

* Author to whom correspondence should be addressed. Tel: +852 2358 8412. Fax: +852 2358 0054; E-mail: kemckayg@ust.hk.

per year. A more recent report released by the UNEP in July 2009 has shown stunning global facts indicating rapidly growing generation of e-waste.

The adverse environmental and health risks caused by E-Waste are of growing importance. E-wastes are made of a variety of organic substances and metals in a polymer matrix. The organic substances, mainly fire retardants and other additives are potential carcinogens. They accumulate in the food chain and the environment due to their non-biodegradable nature. Toxic heavy metals such as cadmium, chromium, mercury and lead are released to the environment mainly through leaching and incineration of the e-wastes. These metals are highly toxic and pose high risk to human health and are a major environment impact. In addition to the toxicity of the e-waste materials, the production process of electronics is often a very energy intensive endeavor. For example, the production of a single personal computer and monitor takes at least 240 kg of fossil fuels, 22 kg of chemicals, and 1.5 tons of water. This amount of water alone, is more than the weight of a modern automobile. This situation further increases the burden to the environment.

Having realized the hazards of E-waste, many developed countries are taking actions. In January 2003, The European Union Council addressed the serious issue of electronic waste streams by the Waste Electrical and Electronic Equipment (WEEE) Directive to motivate the reuse, recycling and reduction of WEEE by efficient material recovery methods. It also seeks to minimize the environmental impacts of WEEE by conducting life cycle assessments. Another regulation, the Restriction of Hazardous Substances (RoHS) Directive, has also become EU law in February 2003. It has been established to restrict the use of hazardous substances in WEEE aiming to reduce its impacts on human health as well as the environment. Followed by these EU's pioneer actions, other countries such as Canada, the United States, and the UK, have developed (or partially developed) e-waste related laws and directives in recent years.

Although the developed countries attempted to reduce the amount of WEEE by regulation, huge quantities of their WEEE are still exported to developing countries including China, Pakistan, India and Indonesia through different channels. The low concentrations of precious metals in e-wastes provide enough economic incentive for less affluent countries to attempt resource recovery methods. However, most of their e-waste treatment facilities and metal recovery processes are primitive which create significant pollution to the air, water and land resources. The local people in turn suffer from a variety of syndromes due to over-exposure to the toxic substances. Thus, the technological advances in the separation and purification of metals and even organic substances from e-wastes are crucial for minimizing the effects on the environment in the recovery processes.

To tackle these challenges from the increasing production and disposal of e-waste, more actions should be taken by the governments including arousing the public awareness, developing environmentally friendly separation and purification processes for recovering valuable materials from e-wastes, establishing new environmental policy and laws, providing safety measures for workers in the recovery facilities, etc. In all probability, e-waste will sooner or later become a commodity due to the abundance of limiting resources such as gold, silver, indium and other precious metals in these solid wastes.

What we suggest is the concept of "mining from the e-wastes". With advanced separation technologies, the processes can be developed to take full advantage of e-wastes in an environmentally friendly manner. Here, a sustainable and feasible treatment process for printed circuit boards is proposed and described. Its applicability for metal recovery from e-waste is discussed.

1. INTRODUCTION

1.1. General Situation

Electronic wastes or e-waste is a generic term for any form of Waste Electric and Electronic Equipment (WEEE) consisting of microprocessors or transistors for digital processing of any form of data. Although this broad definition can be used for the diverse electronic products emerging in the market place today, it introduces difficulty for knowing the actual compositions of the e-wastes which is essential for further development of environmental conservation, recovery and recycling. Since 2002, the Directive 2002/96/EC of the European Parliament and of the Council [1] provides a more rigid definition e-waste (EU, 2002a). Waste Electrical and Electronic Equipment, defined in the directive which classifies e-wastes into ten categories as listed in Table 1.

Table 1. WEEE categories according to the EU directive on WEEE [1]

Number	Category
1	Large household appliances
2	Small household appliances
3	IT and telecommunications equipment
4	Consumer equipment
5	Lighting equipment
6	Electrical and electronic tools (with the exception of large-scale stationary industrial tools)
7	Toys, leisure and sports equipment
8	Medical devices (with the exception of all implanted and infected products)
9	Monitoring and control instruments
10	Automatic dispensers

With advancements in the electronic world almost occurring on a day-to-day basis and increased availability of products to the public, it is not surprising to see a staggering increase of electronic wastes over the past several decades. According to a United Nations Environment Program (UNEP) alert bulletin in 2005, 20 to 50 million tonnes of e-wastes are produced every year with an estimated annual increasing rate of 4% [2]. Miniaturization of electronic devices such as mobile phones due to technological advancement does not slow down the growth as the effect has been counter-balanced by increased sales, particularly in developing countries.

Indeed, the production of electrical and electronic devices is expected to be the fastest-growing sector in industrialized countries even with implemented environmental regulations based on a simulation result [3] - The total mass flow of WEEE increases by a factor of 2.7-4.0 even within an environmental regulatory system while a factor of 3.1-7.0 is expected without the regulation. The fast growth rate is due to the economic growth in the industry [4], shorter lifespans of electronic devices [5] and the development of diversified electronic related products in recent years [6].

Several statistical ways to quantitatively estimate the generation of WEEE have been adopted worldwide [7]. A Swiss study estimated the value based on an assumption that the electronic equipment market is already saturated so that any newly implemented equipment would imply the disposal of the old equipment to be replaced [7].

However, this method does not account for the re-use and storage of the old equipment. In Germany, the production of WEEE is assumed to be equivalent to the production and sales of electronic products. In some other countries, the average number of electronic equipment owned by a typical household is estimated. By knowing the populations and service life of the equipments, the amount of WEEE is determined. Comprehensive summaries of the global production of e-wastes were reported by Robinson [8] and Gaidajis et al. [9].

As WEEE represents a huge variety of products, the exact chemical compositions are unable to be determined. In general, e-wastes consist of more than a thousand different substances [10] and these substances can be further categorized into metals and non-metals. For instance, the metal content of a desktop computer is listed in Table 2.

In addition to these metals, the desktop computer also contains plastic and glass in which other constituents are embedded such as fire retardants and other additives.

Table 2. Metal content in a typical desktop computer[1] [11]

Metal	Content (%)	Weight (g)
Aluminum	14.17	3779
Cadmium	0.0094	2.507
Copper	7	1847
Chromium	0.0063	1.68
Gold	0.0016	0.43
Iron	20.47	5458
Lead	6	1679
Nickel	0.8503	226.7
Mercury	0.0022	0.5867
Silver	0.0189	5.04
Zinc	2.204	587.9
Others	52.12	314.8

[1] The values were base on a typical desktop computer weighing 27.3 kg.

Although only very limited amounts of most of these elements exist in each portable unit, the total quantity of the wastes is actually large due to the huge number of the devices being disposed. Puckett and Smith [12] reported that the disposal of 500 million PCs between 1994 and 2003 contain approximately 2,872,000 tonnes of plastics, 718,000 tonnes of lead, 1363 tonnes of cadmium and 287 tonnes of mercury. Another example is the quantity of disposed mobile phones. More than a billion mobile phones are produced annually worldwide. Manufacturing mobile phones and personal computers consumes 3% of the gold, 2.5% of the silver, and 12% of the palladium mined annually world-wide. 15% of cobalt mined annually worldwide is also incorporated into these devices mainly as a component of the Li-ion battery [13]. A more accurate estimation of the electronic waste composition has been conducted by the measurement of actual samples collected on-site [14].

To minimize the amount of WEEE being generated, governments of developed countries have established new environmental regulations and guidelines to tackle this problem.

One of the approaches is to force or encourage the manufacturers to recycle their products after their service-life. This is known as the extended producer responsibility (EPR) practice promoted by the Organisation for Economic Co-operation and Development (OECD) [15]. Table 3 lists several examples of EPR practice. Recently, the European Community established the Restriction of Hazardous Substances Directive (RoHS) [16] which restricts the use of certain hazardous substances in electrical and electronic equipment. The six substances include lead, mercury, cadmium, hexavalent chromium, polybrominated biphenyls (PBB) and polybrominated diphenyl ether (PBDE).

Table 3. Possible approaches to EPR and examples [15]

Type of EPR approach	Examples
Product take-back programs	Mandatory take-back
	Voluntary or negotiated take-back programs
Regulatory approaches	Minimum product standards
	Prohibitions of certain hazardous materials or products
	Disposal bans
	Mandated recycling
Voluntary industry practices	Voluntary codes of practice
	Public/private partnerships
	Leasing and "servicizing"
	Labelling
Economic instruments	Deposit-refund schemes
	Advance recycling fees
	Fees on disposal
	Material tax/subsidies

Nevertheless, treatment of WEEE after the service-life of the electronic products is inevitable and the conventional solid waste management including landfill and incineration are not wise options for treating WEEE due to the existences of toxic chemicals such as heavy metals and halogenated organic compounds. Landfill leachates will contain toxic compounds and incineration emissions will contain dioxin related compounds and volatile heavy metals. The adverse environmental and health risks caused by WEEE are of growing importance. Unfortunately, instead of treating the wastes locally, more than 50-80% of the WEEE generated in most developed countries is exported to developing countries such as China, India, Pakistan and Africa [17, 18]. This is done despite the fact that long distance transportation associated with e-wastes increase the overall environmental burden based on a recent life-cycle assessment study [19]. To prevent serious damage to the environment, the 1989 Basel Convention [20], signed and ratified by countries over the world except Afghanistan, Haiti and the United States, provides a guideline for the exporting countries to ensure that hazardous wastes are managed in an environmentally sound manner in the importing countries.

1.2. Situation in Hong Kong

In Hong Kong, the amount of WEEE generated in the city has been increasing at a steady rate of approximately 2% per year for the last several years based on the official statistics provided by the Environmental Protection Department in Hong Kong. The amount of WEEE production per person is about 10 kg per year. This figure is higher than the United States estimated at about 7 kg per person per year in 2005 [21]. For South Africa, Brazil and India this number is approximately 1.3, 2, and 0.4, kg per person per year respectively [4, 22].

To tackle this problem, the government of Hong Kong has initiated an undertaking for the emerging e-waste problem. Since 2003 there have been voluntary WEEE collection and recycling schemes which have been successful. These pilot programs processed 60,000 units of computers and electrical household appliances in 2008. Currently the voluntary governmental programs collect close to 90,000 units a year. However, this number accounts for approximately only 1% of the annual WEEE generated in Hong Kong; thus the current programs are inadequate to meet the territory's needs. More programs and processing plants are underway, such as a pilot WEEE processing facility at Kowloon Bay with an annual capacity of 130 tons, and the EcoPark project with an annual capacity of 250 tons. Nevertheless, compared to the huge amount of E-waste produced in Hong Kong, many more undertakings are required.

In a recent Consultation Document, a New Producer Responsibility Scheme has been drafted [23]. According to statistics, 80% of the electronic wastes in Hong Kong are recycled; this number is much higher than the global average. However, this figure should be approached with caution as most of this recycling is in the form of selling the WEEE to developing countries through second-hand dealers. The remaining 20% percent is sent to landfills. As far as the economics of such undertakings are concerned, the New Producer Responsibility Scheme states that since most of the WEEE is exported out of Hong Kong, the remaining amount is not enough to justify a viable commercial operation. The capital investment would be too high in such a situation. So unless cheaper and more holistic processes could be developed, and also, more WEEE is kept in Hong Kong, such projects are not at the moment economically attractive.

1.3. Printed Circuit Boards

WEEE exists as a complex mixture of chemicals, including both organic and inorganic substances. To recover the trace amount of precious metals, the treatment process of WEEE involves a series of physical operations and chemical reactions. During the process, the toxic constituents present in WEEE are released to the environment, causing pollution. A number of pollutants have been identified in different kinds of WEEE. For instance, numerous polymer additives such as fillers, colors, antioxidants, flame retardants, stabilizers, plasticisizers, blowing agents and surfactants are found in the non-metal fraction of e-wastes [24]. Many different commercially available halogenated compounds can be found as flame retardants [25]. As the compositions of e-wastes vary with its categories, thus, it is almost impossible to identify or monitor the exact chemical compositions of the general waste stream, despite the effort on development of more advanced analytical techniques [26]. To limit the context in this report, only printed circuit boards are reviewed; as they contain most of the precious metals in e-wastes.

Printed Circuit Boards (PCB) are in fact the generalized term representing the platform upon which microelectronic components such as semiconductor chips and capacitors are mounted. They are used to support the electronic components as well as to connect them using conductive pathways, tracks or signal traces etched from copper sheets laminated onto them. In literature, a PCB is also referred to as a printed wiring board (PWB) or etched wiring board. A PCB populated with electronic components is a printed circuit assembly (PCA), also known as a printed circuit board assembly (PCBA). In this study the term PCB is used in place of PCBA unless otherwise stated.

For the manufacturing of a PCB, conducting layers of thin copper foil and insulating dielectric composite fibers are used. This copper can be applied to either one, or both sides of the substrate. Then, the unwanted copper is removed, leaving only the desired copper traces. A minority of boards are produced by directly adding traces of copper to the bare surface. The amount of copper used is in relation with the current that the conductor must carry. Signal traces are typically narrower than power or ground traces. This is because the current requirements of signal traces are typically much smaller. In a board with several layers, if one layer is to be designated as the ground plane for shielding and power return, it might consist of an entire layer of solid copper.

A major issue arising from the advancements in the speed and functionality of components used on PCBs, is the availability of materials for the PCB substrate that are compatible with these products and their process needs. This includes the stresses created by higher temperatures during the assembly process. In addition, the expansion of the components and the substrate due to heat must be considered [27]. The National Electrical Manufacturers Association (NEMA) has classified the grade material of these composites as seen in Table 4. It should be reminded that NEMA grades represent only material categories and not one individual material. For example, the very common FR-4 is not a specific laminate material, rather it is a class of material with similar properties [28]. For computers and communication equipment, FR-4 is most common, while for home electronics and television, FR-2 is predominantly used. Today, high-value equipment increasingly contains FR-4 boards [29].

For temperature performance higher than that of FR-4, other resins such as polyimides, cyanates, PTFE, other fluoropolymers, epoxy-PPE blends, and even ceramics are often used [30]. G-10 is usually not used for most standard PCBs, with its main usage being for thin watch circuits due to the fact that it is water resistant and can be punched easily [31].

Due to the risk that circuit boards might ignite at high temperatures in the processing of the components and connections (i.e. during the soldering process), and also due to flammability risks as the result of electric energy impacts, materials with low flammability are required for the production of PCBs. Traditionally brominated fire retardants have been the most important and popular fire retardants in this sector. Although phosphorous or nitrogen based fire retardants are available as an alternative and post-industrialized countries are gradually moving towards ending halogenated fire retardants in PCBs, brominated fire retardants are still overwhelmingly predominant on a global scale [32].

Table 4. NEMA Grade Descriptions and Properties [30]

NEMA grade	Resin, Reinforcement	Description	Typical uses
XXP, XXXPC	Phenolic, paper	Hot or cool punching	Inexpensive consumer items like calculators
FR-1 and FR-2	Phenolic, paper	Flame retardant XXP/XXXPC	Where flame retardancy is required
FR-3	Epoxy resin, paper	Flame retardant	High insulation resistance
CEM-1, -3	Epoxy, glass cloth, paper or glass core	Punchable epoxy, properties between XXXPC and FR-4	Radios, smoke alarms, lower cost than FR-4
G-10	Epoxy, paper, glass	Excellent electricals, water-resistant, not F.R.	Computers, telecom, cost above FR-2
FR-4	Epoxy, glass cloth	Like G-10 but flame retardant, Tg ~ 130oC	Where flame retardancy is required
FR-5	Epoxy, glass cloth	Like FR-4, more heat resistant, higher Tg than FR-4	Military, aerospace, where specified
FR-6	Polyester resin, glass mat	Flame resistant	Low capacitance, or high impact apps.

In order to populate the board, components must be attached by electrically and mechanically fixing them on with a molten metal solder. In recent decades the solder was most often a tin-lead alloy. However, with environmental legislation in the EU, Japan and the USA restricting the use of lead, many new solder compounds have been developed [33-35]. Since PCBs are manufactured in various types and sizes, ranging from single-layered to multi-layered, and single sided to double sided, and that the components placed on each PCB can be variant both in function and in material, any "average" given of the constituent materials or of the size and weight of PCBs must be approached with caution. Average values are often very much dependent on the boards under study. The disparity of material concentrations varies from board to board. Nevertheless, there have been several studies conducted to obtain average values of material compositions. Most of these studies divide PCB based on their functions (i.e. computers, phones, etc.) and in turn, provide averages for these subgroups [36-39]. The challenges and opportunities from the imported e-wastes, particularly printed circuit board assemblies, for developing countries are discussed in the following sections.

2. Challenges

2.1. Unknown Quantity of WEEE

As previously discussed, developed countries established various regulations and guidelines to minimize the environmental impacts of WEEE disposal recently. However, most of the e-waste is exported to the developing countries through semi-official or unofficial ways and no official statistical data shows the actual quantity being exported. Indeed, it can only be estimated by a preliminary model [40]. On the other hand, due to the rapid economic growth in many developing countries, the domestic generation of electronic wastes in these countries is also expected to increase drastically; such as the situation in India [41], Nigeria [42] and China [43]. Since the WEEE contains various toxic chemicals in its matrix, the import of e-waste by less affluent and developing countries, is not very much different from the import of toxic wastes [17].

2.2. Lack of Technological Advancement

As the technological levels of these countries are generally low, the infrastructures and facilities for proper treatment of WEEE are absent. Hence, huge quantities of the toxic wastes are mostly treated by primitive processes, namely manual dismantling, open burning and acid leaching [44]. Various toxic pollutants, both organic and inorganic, are released to the environment by these primitive technologies. Polycyclic aromatic hydrocarbons [45], halogenated compounds such as flame retardants [46, 47], polychlorinated biphenyls (PCBs) [47] and other combustion products during the thermal degradation of the non-metal fraction of e-waste are also released to the environment [48, 49]. In addition, heavy metals such as lead, cadmium, mercury, copper and chromium are leached to the environment [50, 51], causing serious air [52], water [53] and land pollution [54-56] simultaneously. Based on various studies, the pollutants are mainly generated and distributed widely by open site burning.

2.3. Lack of Safety and Health Awareness

Although the local citizens benefit from the precious metal recovery, the environment and health of their offsprings most likely suffer from the substantial release of the toxic components formed in the recycling processes. Sepulveda et al. [57] and Leung et al. [58] conducted reviews of the environmental fate and effects of hazardous substances released from electrical and electronic equipments during recycling in China and India. The concentration levels of various pollutants are found to be significantly higher than the reference values suggested by the World Health Organization (WHO). These contaminants enter the food chains in the affected areas and finally accumulate in human bodies by ingestion or inhalation routes [59]. Unfortunately it has been documented that these pollutants have been found in the bodies of infants and children as well [60-62]. Damage to DNA and chromosomes [63] and disturbance to hormone levels [64] are the identified adverse effects of these pollutants on human health. Despite the fact that the primitive treatments of the e-wastes are polluting, the workers and residents may minimize the effects by implementing safety measures such as personal protective equipments. However, since the citizens and workers are generally inadequately educated, they lack awareness and do not know of the serious risks concerning their safety and health.

3. Opportunity

As previously mentioned, e-wastes do not only contain toxic substances, but they also consist of quantities of valuable materials such as metals. By extracting these valuable materials from e-wastes, huge profit can be generated. Thus, e-waste can also be recognized as a kind of resource if a technical, economically feasible, and environmentally friendly process is available for recovery. According to The Encyclopedia of Earth, one metric ton of circuit boards can contain between 80 and 1,500 g of gold and between 160 and 210 kg of copper. To put this number into context, it should be stated that these concentrations are 40 to 800 times the amount of gold in gold ore, and 30 to 40 times the concentration of copper in copper ore mined in the United States [65]. Therefore, the Printed Circuit Boards which house these metalsrepresent the most economically attractive portion of WEEE. Yet, the fact that such a highly complex concoction of various valuable and sometimes hazardous materials are intermingled in such a small volume, poses serious engineering challenges for the recovery and recycling of the constituent materials. The heterogeneous mix of organics, metals, fiber glass, toxic materials including heavy metals, and plastics make processing PCBs a challenging task [66]. In addition, the average rate of PCB production has been increasing more than 8 percent annually in past years. For Southeast Asia this number is higher than 10 percent, and in China, nearly 14.5 percent. In 2005, the industry passed the 10 billion dollar mark in mainland China, and it was projected to pass the 12 billion dollar mark by 2006. Such growth adds incentive to finding methods of effectively recycling and mining the valuable constituents of these boards.

New technological advancements in today's world have lead to an ever increasing pace at which electronics become obsolete. The acceleration of the replacement of equipment has thus led –and will lead to- a significant increase of waste printed circuit boards [67]. These waste printed circuit boards can be viewed as both a secondary resource of materials, and a contaminating environmental burden. It was also stated previously that due to the heterogeneous concoction of organic materials, fibers and metals, these wastes are particularly hard to recycle and treat. This demonstrates the necessity to develop treatment methods in order to both take advantage of the economical prospects, and also, safely treat the hazardous materials within printed circuit boards.

There have been many studies focused on particular aspects of this treatment and metal recovery. These processes include pyrometallurgy methods [68], hydrometallurgy methods [69, 70], mechanical/physical processes [71-76] and thermal processes [77-81] alongside others.

The problem with these studies is that they often focus on very particular portions of the entire process [82], laboratory scale experiments, and only the treatment of a specific material. A wide range of studies exist in specific treatment methods. Such as physical beneficiation techniques [83], the use of supercritical methanol for metal recovery [84], and separation processes using various mills such as stamp mills, hammer mills and pin mills [85-87]. Researchers have even gone as far as trying to produce nano-Cu_2O/TiO_2 photocatalysts from waste PCBs to some degree of success [88]. The aim of the following sections is to investigate the hundreds of studies conducted in this regard and collate together the numerous studies conducted in the field of PCB treatment to devise an integrated process for treating circuit board assemblies.

3.1. Disassembly

Before PCBs can be processed they must be disassembled and separated from the WEEE. Based on the fact that WEEE comes in various different shapes and sizes, it is impossible to develop a mechanized process for disassembling PCBs from all WEEE. Thus at present, manual disassembly is the most practical option. For the sake of this project, it is assumed that waste PCBs are available as the raw material to be treated.

3.2. Component Removal

There are several research projects underway for developing automated and semi automated processes for removing the different components from PCBs [89].

Our proposed process first begins with the removal of solder from the PCBs. It should be noted that the components of circuit board assemblies are held on to the board via solder; so effective recovery of solder would in turn lead to the separation of the components from the board itself.

Electrochemical methods have been proposed for removing the solder [90]. However, thermal methods which are much easier and have less environmental impact seem like a more logical choice. For example in one proposed study [91], heat sources such as infrared radiators are placed on both sides of the board which is held in a horizontal position. Heating occurs and when the solder melts, the hanging components will fall off. The components on the surface in turn, are removed by a brush.

The thermal process chosen for this report is as follows [92]: The waste PCBs are submerged in a drum containing diesel oil. The drum is punctured evenly with holes on its side and bottom. Then the diesel oil is heated until the solder melts. The drum is rotated with an electromotor and as a result, the molten solder seeps through the holes and falls to the bottom of the reactor.

The oil temperature is an important factor which must be monitored and managed. The higher the temperature, the less viscous the molten solder will be and the easier its removal. However, if the temperature is too high, the board will be damaged and some toxic substances will be released. Most conventional solders used in PCBs (made of tin and lead) have melting points between 180°C and 190°C. New lead-free solders have melting points 5-20°C higher.

It is found that the board will start to break down and release toxic materials at temperature higher than 240°C. So the temperature must be kept between this window – as close to 240°C as possible. At such temperature with a rotation of 1400 rpm for 6 min intermittently, the solder is completely removed; all the components are separated, and due to the oil surrounding the board, the chemical properties of the recovered solder remain unchanged rendering it "good-as-new" for direct reuse.

Another extremely important advantage of removing the solder is that the constituents of solder (mainly tin and lead) will not pose further problems in the remaining treatment process steps. The process schematic for solder separation is displayed in Figure 1.

From here, the solder needs no further processing and is ready for reuse. The component-less boards and the components move on to further processing steps.

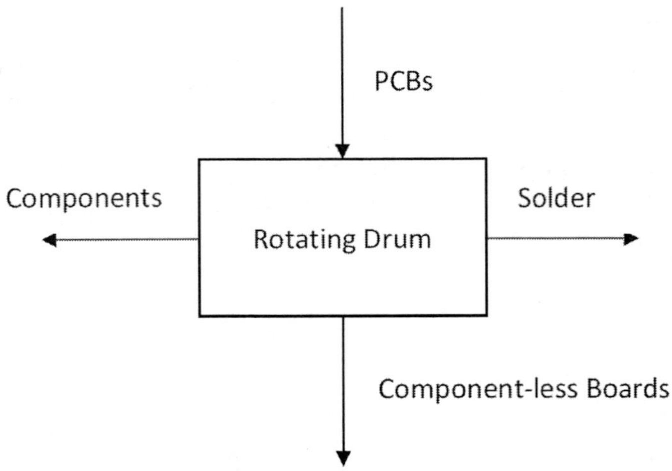

Figure 1. Schematic of solder separation.

3.3. Pyrolysis

Traditionally the recycling technologies of PCBs are based upon particle recycling, chemical recycling, incineration and pyrolysis. Each of which have their advantages and their drawbacks.

Particle recycling is to grind the non metal fraction of the boards into fine sizes in order to use them as fillers and in new composites [93-96]. The main disadvantage of this choice is that the fine particles must be rather clean and pure; this is while nonmetal fractions of PCBs are very often contaminated with various contaminants.

The second option of chemical recycling is in fact converting the resin wastes into their original constituents. However, since the boards are a mixture of various types of resins – including epoxy and phenolic resins- these chemical processes are not practical. Chemical processes are usually specific to particular polymer types, are expensive, and do not yield high-quality results [97, 98]. Also, chemical processes very often result in high amounts of polluted effluents which must in turn be processed.

Incineration or combustion recycles the calorific value of the resins and nonmetallic fractions of the boards. However, much of the board is converted into useless slag. This slag reduces the combustion efficiency, and is ultimately to be disposed of in a landfill [99, 100]. If a blast furnace is used, the copper in the board will be reduced to black copper [101]. Also, there are serious air pollution problems which arise from combustion methods such as the creation of harmful dioxins and furans.

Pyrolysis is the chemical decomposition of substances by heating that occurs in oxygen-less environments. In comparison with incineration, the production of harmful furans and dioxins are much less [102], the operation temperatures are lower, the volume of the exhaust gas is reduced [103], and leads to the formation of gases with reasonable calorific values which can be used to heat the process, oils which can be used as chemical feedstocks or fuels, and chars which contain untarnished copper [104]. Still, the solid products include char with little or no value, and the problem of dioxin formation persists [105].

Dioxins have been studied extensively and have been found in the exhaust of thermal decomposition of circuit boards due to the presence of brominated flame retardants used in PCB manufacturing. These compounds build up (bioaccumulate) mainly in fatty tissues and have rather long half-lifes. In 1994, the US EPA reported dioxins as probable carcinogens. In addition, the EPA also stated that non-cancer effects such as adverse effects on reproduction, sexual development and the immune system may pose an even greater threat to human health. Many of the products of the thermal breakdown of circuit boards are seen in the WHO list of toxic chemicals [106]. Although at low concentrations they might not even be discerned by practicing physicians [107], at high concentrations the worst dioxins can be considered "next to the nuclear catastrophes" [108].

In order to resolve the problem of dioxin and furan formation various solutions have been proposed – some of which will be addressed here. Generally for feedstock recycling of thermoset resins, processes such as depolymerization with the addition of hydrogen donor solvent [109, 110], depolymerization using supercritical CO_2 [111], pyrolysis alongside limestone [112], multi-stage pyrolysis [113], gasification and bromine recovery via wet alkaline scrubbing [114] catalytic debromation and hydrogenation [115-119], and co-pyrolysis with inorganic solids such as zeolites and sodium hydroxide [120] have been proposed. However, it is very important to note that these proposed processes are mainly aimed at general halogenated plastic mixtures and municipal solid waste. In fact, the studies that particularly focus on printed circuit boards are much less extensive and at times, very limited.

It has been proposed that the resulting oils of the pyrolysis process can be further treated in the presence of polypropylene as a reductive medium [121]. Through the detoxification procedure, molecules such as dioxins are broken down and hydrogen bromine and small brominated alkanes are formed. The optimal operating conditions for the process are found to be 1 Mpa, 350°C, and with a residence time of 20 mins. The process was also tested at atmospheric pressures. However, the reaction between polypropylene and brominated compounds under atmospheric pressures is much less. Unfortunately, although this study has been conducted in 2005, there are no studies which have expanded and further investigated the feasibility of such a process.

A separate study has used CaO to react with HBr and form CaBr2 in the solid phase. However, this study requires a much higher temperature which make it not as feasible and effective as the abovementioned study [99]. Two other Chinese studies (not available in English) have also reaffirmed the effectiveness of calcium carbonate (CaCO3) for debromination – claiming: "the results showed that 93.4% of the inorganic bromides were trapped when the wPCB and CaCO3 blended in the same quantity" [122, 123].

Studies also exist which claim that increasing the temperature will reduce or even eliminate dioxin formation. Some claim not to have detected any dioxin formation at 800°C [81, 102], while another study [124] reports the production of dioxins at temperatures even as high as 1200°C. It seems very unlikely for dioxins to exist at such high temperatures; nevertheless, when such studies exist, one should act cautiously. It should also be noted that increasing the heating rate of pyrolysis results in an increase of the fraction of bromine evolved in gaseous products and a decrease of the fraction of bromine present in the condensable liquid phase [102].

Overall, for the integrated process of this report, the option of using calcium carbonate is chosen. This is due to the fact that the temperature can be kept love thus reducing operation costs. In addition, the calcium bromide (Price: US $1200-5000 / Ton) [125, 126] which is produced as a result of the process has a higher added value than the calcium carbonate (Price: US $50-600 / Ton) [104, 105] used to produce it. So aside from the option of recovering and selling the bromine separately, calcium bromide itself can also be sold with approximately twice to ten times added value (based on the quality etc.).

The block diagram of the pyrolysis process is shown in Figure 2.

Note that the Calcium Carbonate and the board pieces are added prior to entering the reactor in order to provide mixing before entry into the chamber.

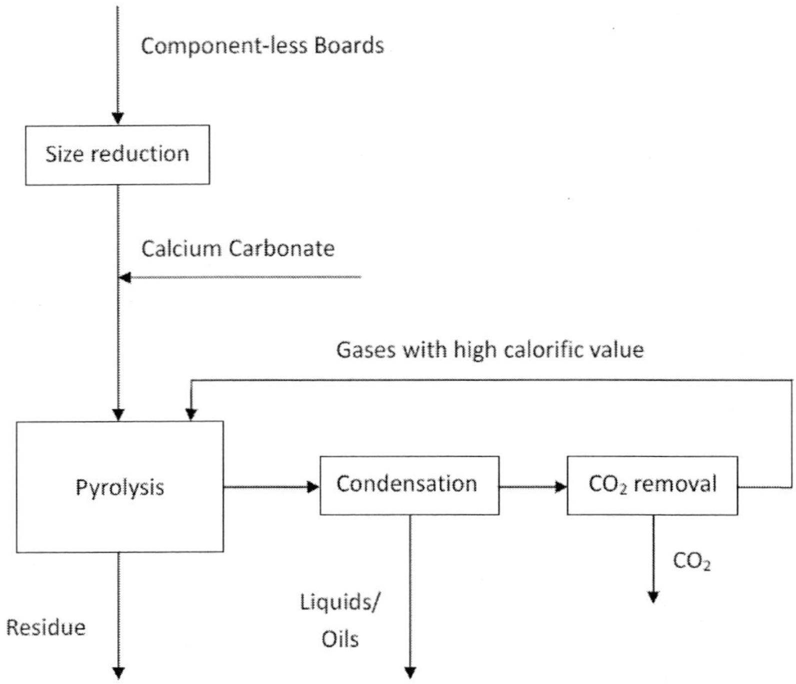

Figure 2. Process flow of pyrolysis.

As it can be seen, the gaseous phase resulting from the pyrolysis is further condensed to produce liquids. The uncondensed fraction of the gas will proceed to CO2 removal and will then be recycled back to the pyrolysis chamber due to its high calorific value [81].

When it comes to the composition of pyrolysis gases, most studies agree that the constituents are usually quite simple. The composition of the gas seems to be dependent on the types of PCB pyrolized, the temperature, and other operating conditions. However the difference is not too substantial.

There are no studies available to show the exact gas composition of the chosen pyrolysis conditions of the developed process in this report. The existence of calcium carbonate will surely have an effect on the composition of the resulting products in all phases. As an example, the amount of water produced when $CaCO_3$ is added was reported to be much higher than any other study without the addition of this substance [123]. The obvious reason of alterations in the resulting compositions, is that the bromine is absorbed by the $CaCO_3$ and the CO_3 is replaced. The carbonate in turn, undergoes alterations and reactions. For the gases generated during pyrolysis, the Gross Calorific Value is reported to approximately be 30

MJ/m^3 [129] which is higher than the amount needed to sustain the needs of the entire process [130, 131]. This calorific value is higher than that of natural gas, and could in some cases even produce excess energy which could be used for other parts of the process [132]. This high calorific value can be understood when considering the composition of the gas on a volume percentage basis. As it can be seen, the biggest volume portion belongs to Hydrogen. Table 5 shows the typical composition of the gas stream from PCB pyrolysis [80].

Table 5. Composition of the pyrolysis gas

Gas	Fraction (vol%)
H_2	38.52
CO	7.23
CO_2	7.93
CH_4	9.55
C_1-C_4	15.23
O_2	12.03
N_2	9.5

As far as the operating conditions of PCB decomposition are concerned, 773 K and 1 atm are adequate. Studies show that most of the chemical breakdown of the PCB occurs prior to 773 K. However, to ensure nearly complete breakdown, 823 K (550°C) is chosen as the designated operating condition for the proposed process in this report.

3.4. Copper Recovery

The residues exiting the pyrolysis chamber must be further processed in order to recover the copper present within them. Before this can happen, the residue pieces must be broken down and shredded into smaller pieces. The heat treatment in the pyrolysis reactor noticeably diminishes the hardness and tenacity of the pieces. Hence, a shear-crusher will now suffice [67]. If these previous treatment steps were not in effect, a more extensive two-step crushing method would be required to achieve desirable size reduction [66]. Nevertheless, after crushing, the pieces will be sieved, and particles greater than 1.2 mm in diameter –if any- will be fed into the crusher again.

It should be noted that there are speculations that copper acts as a catalyst for dioxin formation when flame-retardants are incinerated. [133]. If copper is in fact acting as a catalyst, it would be more reasonable to separate the copper before the pyrolysis chamber. This would be a suitable tradeoff, resulting in a more extensive shredding process, in return for a much cleaner process with less hazardous problems.

There are many methods present in literature regarding copper recovery. This is most probably due to the fact that copper comprises the largest percentage of any metals on PCBs and therefore has been a more obvious target of research. These include chemical leaching [134, 135], electrostatic and corona separation methods [136, 137] and electrometallurgy processes [138, 139] amongst others [140].

Bioleaching methods have also been recently studied. It has been revealed that copper leaching via A. ferrooxidans are completely viable [141]. This method has great potential as it is not harmful for the environment, is cheap, and relatively fast compared to many other biological processes. A. ferrooxidans are first cultivated for 3-4 days as stock culture, until the concentration of Fe^{3+} ions has reached a desired amount considered strong enough to dissolve the metallic copper. From there, the bioleaching will be initiated and under certain conditions up to 100% copper mobilization can be reached within 24 hours.

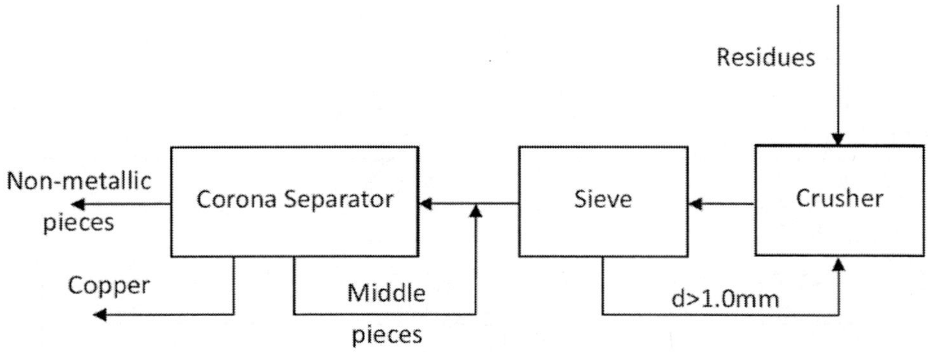

Figure 3. Process flow for copper recovery.

For the current process design however, a corona separation method will suffice. This is because the other components have been previously removed, and since the remaining residue effectively has only one metal component –mainly copper- the corona separation would be quick and simple. Also, when compared to other physical methods, corona electrostatic separation has high productivity, no wastewater or exhaust gas, and low energy usage. It has been reported that this method requires about 5% of the energy that would otherwise be required compared to a pneumatic separator [66].

During the separation process, metal particles fall into the first container while the non-metals continue along the process and fall in last container. The middle container will collect less than 10 percent of the particles which can in turn be fed into the separator once more. Almost all the copper can be recovered with this method at a productivity speed of 0.5-1.0 ton/h for each corona electrostatic separator. From there, the copper can be directly sent to the smelting plant. The block diagram for copper separation is shown in Figure 3.

3.5. Treatment of Liquid Products

The liquid products that result from pyrolysis are mainly composed of phenols. The composition of the oil will be directly dependent on the pyrolysis conditions, the PCB composition, and the effect of the added Calcium Carbonate. It should also be noted that there is a percentage of water present in the liquid phase which can be separated from the oils.

The calorific value of the liquids resulting from PCB pyrolysis has been reported at approximately 26.5 MJ / kg [129]. Thus, the liquids can also be classified as fuel grade material. In comparison, the calorific value of petrol, ethanol, and wood are 43, 30, and 16 MJ/kg respectively [141].

The problem with the separation of the various components of the oils is that due to their variety, the purification will be costly and complicated. As a result, for the proposed integrated treatment process of this study, the synthesis of carbon nanotubes and porous carbon from these oils is chosen. In a recent study published in 2010, the mechanisms of both these processes are explained [142]. The possibility and feasibility of using pyrolysis oil as a precursor for advanced carbonaceous material (ACM) is acknowledged. Since PCB pyrolysis oils contain a large percentage of phenols and phenol derivates, such processes are applicable. First the oils are polymerized in formaldehyde solution to synthesize oil-based resin which is in turn used for the production of either CNT or porous carbon. The authors claim that the quality of the advanced carbonaceous materials produced in the study is satisfactory and that their production from pyrolysis oil can compete with industrial processes due to the simple processing method and cheap price [142]. As a suggestion for further studies, we recommend the investigation of whether dioxins are destroyed in such a process or not. Due to the various thermal and chemical treatment steps with temperatures reaching as high as 700 and 900°C, the production of thse advanced carboneous products might automatically do away with the dioxins in the liquid phase. If so, the cumbersome preventive measures taken in order to reduce dioxin formation could possibly be bypassed. This would in fact, be an extremely important accomplishment in the path of PCB pyrolysis and would make this processing route much more attractive. Figure 4 shows the process flow for treating liquid products from pyrolysis.

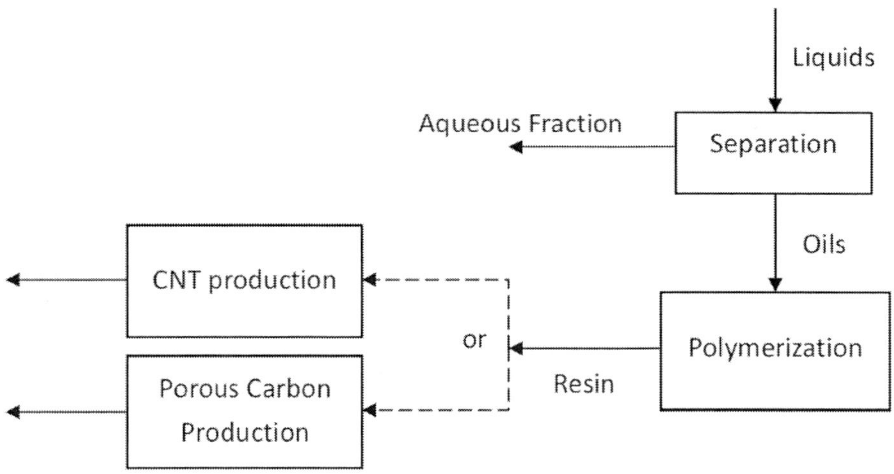

Figure 4. Process flow for treatment of liquid products from pyrolysis.

As previously stated, many of the elements in the periodic table can be found in printed circuit boards. Most of these elements are within the various components placed on PCBs. However, the concentrations of these elements are so small that their recovery is not feasible.

There are many studies that address the subject of metal recovery from electronic waste; some of which include established pyrometallurgical recovery [143-147], hydrometallurgical processing [148-153], and biometallurgical methods [154-158].

In a review of metallurgical methods for recovering metals from WEEE, Cui et al. [133] compared various options and stated that traditionally pyrometallurgy has been used for this feat but these methods have recently faced challenges due to environmental considerations.

On the other hand, modern smelters depend very much on high capital investment. In the past decade hydrometallurgical processes have been investigated. As far as gold is concerned, after examination of economic and environmental factors, it is concluded that from among the conventional methods the leaching of gold by thiourea is a good option and replacement for cyanide. The review also states that biotechnology is a promising area of research. Biotechnology has been used for many years for the extraction of metals from ores [158], however, the research regarding e-waste and in particular PCBs is extremely limited [159, 160].

For the process under development in this report, in order for the process to proceed and the various metals to be extracted, the components need to be shredded into smaller pieces. After going through the shredder, the pieces are then passed through a sieve. Particles with diameter larger than 0.5mm are separated and fed back into the shredder for further crushing. The shredding process will produce dust which will include a considerable amount of metals [161]. Much of the metals could be lost in this stage. For example, palladium is often found in ceramics that are broken down to dust during shredding; this might be an explanation as to why some studies have shown that the concentrations of precious metals decrease as PCBs are shredded. The limits of technology will simply not allow achieving recoveries of 100% [162].

After the shredding, the iron is removed with magnetic separation. Following the magnetic separation, the collection of particles will be sent to a bio-tank. The bio-tank will contain a mixed culture of T. ferrooxidans and T. thiooxidans. These microorganism will have been previously grown for 7 days in the absence of PCBs in another chamber. The PCB component pieces will be added to the culture for 10 days [155]. As a result close to 100% of the existing Al, Ni, and Zn will mobilize. The remaining PCB particles will move on and be cleansed from the microorganisms.

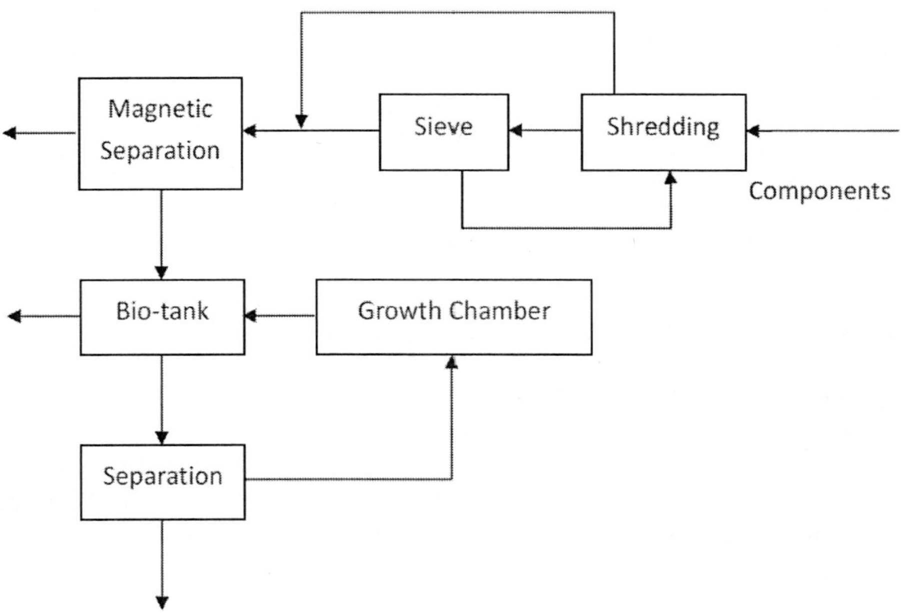

Figure 5. Process flow for precious metal recovery.

After this stage, the remaining PCB particles are mainly comprised of Ag, Au, Pd and non-metallic particles. There are a very limited number of very recent studies which have addressed the possible extraction of these precious metals with bio-methods [159]. Naturally, these studies are at their infantile stages and cannot be relied upon at the moment. Thus, although the development of these studies is highly recommended, for the proposed process, an established hydrometallurgical method will be chosen [163, 164]. An optimized flowsheet at this stage will be able to recover 93% of the silver, 95% of the gold, and 99% of the palladium which enters this stage.

3.6. Case Study in Hong Kong

For the mass balance, we assume that from the 71400 tons of E-waste produced in Hong Kong which yields 2213.4 tons of PCBs (3.1% of WEEE), only 1% is treated in the proposed process plant. Thus the input into the process is 22.134 tons of PCB per year.

In the first step of the process, i.e. the solder removal, around 4% mass of this amount will be removed as solder. This is equivalent to 0.88536 tons. It is noted that no studies exist that the author of this report is aware of which indicates the weight fraction of components on PCB with respect to the total weight. Thus a very approximate assumption is made here. Based on the solder prices provided in 2006, normal solder would cost approximately 6390 US$ per ton; while lead-free solder would cost more than 2.5 times at a price of around 23170 US$ per ton.

It is assumed that the board is comprised of 90% non-metallic fractions plus copper; and the remaining weight belongs to the other components on the board with an additional assumption that the copper in the components is negligible when compared to the wiring on the board. Studies indicate that 70% and 16% of the PCB (including components) is comprised of non-metals and copper respectively.

Since the majority of the solder used in PCBs is in the traditional type, then the gross economic value of this step would be 6390*0.88536 = 5657.45 US$ per year. With the limitation of the use of lead in solders and recent legislation, the future gross income of this stage (whence all solder is lead-free) will be around 20513.79 US$ per year.

The average weight percentage of residues, liquids, and gases produced due to the Pyrolysis of PCBs vary from study to study, and PCB to PCB. Barontini and Cozzani reported 62%, 32% and 6% respectively [102] whereas Zhou and Qiu reported the values to be 69.5%, 27.8% and 2.7%, and 75.7%, 20.0%, and 4.3% respectively [92]. For the estimation here – if no calcium carbonate was added- the weight percentage of 70, 20, and 5 would be designated respectively for residues, oils, and gases. The remaining 5 percent would be the aqueous content of the liquid phase. The products from the proposed pyrolysis process would thus include 12.2402 tons residues, 4.3715 tons liquids (3.4972 tons oil, 0.8743 tons aqueous fluid), and 0.8743 tons gaseous products (approximately 23170 MJ/year).

For the pyrolysis with calcium carbonate, the main difference is that the mass of calcium carbonate which is added to the process will be considered as part of the residue. This is a good assumption since the melting point of $CaCO_3$ is higher than the pyrolysis temperature. The effect of calcium carbonate on the liquid and gas yields and compositions could be an area of future research. Currently bromine content of PCB can be as high as 15-20% weight [165]. However, due to legislation and newer imposed restrictions, more recent publications give figures closer to 5%. Thus, taking the optimum 9 to 1 ratio of Ca to Br proposed in [124], the process would require around 7.9 tons $CaCO_3$.

97% of the copper remaining in the components is proposed to be mobilized by the mixed T. ferrooxidans and T. thiooxidans culture in a bio-tank. So the total copper removal weight can be regarded as 97-100% of the entire copper content within the PCB, which is around 3.49 tons. Taking the price of copper as 7200US$/ton [166], it is equivalent to 25128 US$. For the non-metal fraction, depending on the route to be taken, the yield of products will vary. If Carbon Nanotubes are to be produced, the yield will equal to 56.82 %. If porous carbon is to be produced, the yield would stand at a lower amount of 38.05%. The price of advanced carbonaceous materials is very much dependent on their quality and the market. Prices for nanotubes range from about 1 dollar per gram to thousands of dollars per gram [167, 168].

For the metal recovery, 1.107 tons Al, 0.4427 tons Ni and 0.2213 tons Zn and 0.664 tons Fe can be recovered. Thus, to mobilize these metals, a relatively large mass of material will go through the bio-tank. Since the optimum concentration of scrap within the tank is found to be at 10g/L (132), this mass of untreated metals would be an unbearable burden for the system and would require a huge amount of microorganisms. Thus, it is logically suggested that the non-metal fraction be removed before it initially reaches the bio-tank. For this, a corona separator similar to the one explained in the aforementioned copper recovery section can be used. Hence, a much smaller mass of materials with go through the bio-tank.

Another advantage of separating the nonmetal fraction before the tank, is that this fraction can be recycled and fed into the Pyrolysis chamber for further liquid and gas production.

If all the non-metal fractions were removed, still close to 2 tons of shredded component particles would remain which would have to go through the bio-tank. At the optimum concentration of 10g/L this would require the tank to be around 2 x 10^5 liters. Thus a tank with approximate dimensions of 10 x 10 x 2 meters would be needed. This is in fact a huge tank. If building such a tank has too high of a capital cost, a trade-off could be devised in which the concentration of PCBs in the tank is increased. The concentration of 50 g/L might be more practical.

It should be noted that considerable revenue (about 28%) is lost due to decreased efficiency. Thus in economic considerations, the benefit of the tradeoff should be delicately scrutinized. Nevertheless, the remaining metals which pass through the bio-tank will then move on to the next step of the process. This mass is now comprised of precious metals alongside other materials. About 0.01107 tons Ag, 0.00664 tons Au and 0.002213 tons Pd exist in the remaining residues. In numerous studies it has be constantly stated that gold recovery is the most attractive economic incentive for E-waste treatment. However, if the production of carbon nanotubes are undertaken, this report has shown for the first time that more than half of the revenue generated from PCB treatment would be associated with advanced carbonaceous material production. It should also be pointed out that the developed "practical" process will recover about 60 percent of the maximum theoretic revenue. Some losses are impossible to avoid (such as losses due to shredding) and some have been inflicted due to the use of environmentally friendly methods and/or methods with low capital and operational costs. All the income generated by the developed process is summarized in Table 6.

Table 6. Practical Gross Revenue of developed process for PCB recovery

Material	Mass (kg) practically produced/recovered per ton PCB	Practical Gross Revenue US$ per ton PCB	Percent of Total Practical Gross Revenue
Solder	40.00	288.70	0.92
Cu	157.68	1135.27	3.64
$CaBr_2$	39.66	47.59	0.15
CNT	74.09	18523.54	59.31
Fe	30.00	9.00	0.03
Ni	17.40	352.09	1.13
Zn	9.50	19.38	0.06
Al	2.37	4.82	0.02
Au	0.26	9312.47	29.82
Ag	0.42	266.47	0.85
Pd	0.09	1271.90	4.07
TOTAL	371.47	31231.21	100.00

CONCLUSION

Printed circuit boards pose a serious imminent environmental problem due their increased production and decreasing life-spans. Also, if regarded as sources of materials, they are more concentrated than metal ores found in nature. In this report hundreds of various studies concerning the treatment of waste printed circuit boards were studied so as to develop a suitable integrated process for processing printed circuit board assemblies. From among the research carried out, the most suitable studies were selected and focused on.

A proposed process was developed which employed many different steps. Starting from solder delivery, the components are first removed from the board itself. In turn, the board is then taken to a pyrolysis step, where it will be thermally treated and broken down to form oils and gases. The produced gases are of high calorific value, and are more than enough to fulfill the energy demand the prolysis chamber. The oils are then collected and with a series of thermal and chemical treatments, are developed into advanced carbonaceous materials with extremely high added value. This is in fact the most lucrative part of the process, as it generates more than 50 percent of the entire gross revenue. It is suggested that further research is carried out on the topic of creating advanced carbonaceous material out of the phenol-dominated oil residues of PCB pyrolysis. The residues remaining from the pyrolysis chamber are treated in order to recover the copper. Copper is in fact the biggest metal constituent of all PCBs. Also, calcium carbonate is added to the pyrolysis reactor in order to inhibit the creation of dioxins and furans due to the breakdown of the brominated fire retardants on the PCBs. A fraction of the calcium carbonate is transformed to calcium bromide and can be further recovered as an added value product. The separated components from the board take a different path. They are first shredded and the iron is removed by magnetic separation. Then preferably, the non-metallic fractions are removed and sent back to the pyrolysis reactor. The remaining metals are treated biologically and the nickel, zinc, aluminum, and any of the remaining copper is leached. Bioleaching and other biological processes are only at their infancy. If developed, they could prove to become a grand option for future designs.

From there, the remaining parts move on to hydrometallurgical treatment in order to recover the precious metals. Hydrometallurgical recovery of precious metals is an established process. There are studies of bioleaching processes for this feat; however none of these processes have reached an applicable level of success.

REFERENCES

[1] European Union (EU), Directive 2002/96/EC of the European Parliament and of the Council of 27 January 2003 on waste electrical and electronic equipment (WEEE) — joint declaration of the European Parliament, the Council and the commission relating to article 9. *Official Journal*, L037:0024-39.

[2] United Nations Environment Programme (UNEP), E-waste, the hidden side of IT equipment's manufacturing and use, (2005) Environment Alert Bulletin, Chapter 5.

[3] Hilty, L. M. *Environ. Impact. Asses.,* 2005, 25, 431– 435.

[4] UNEP DTIE Sustainable Consumption and Production Branch, StEP Initiative, Recycling – From E-Waste To Resources, Sustainable Innovation andTechnology Transfer Industrial Sector Studies (2009).

[5] Culver, J. The life cycle of a CPU; 2005. http://www.cpushack.net/life-cycle-of-cp.*u.html.*

[6] O'Connell, K. A. Computing the damage, waste Age; 2002. http://www.wasteage.com/ar/waste_computing_damage/.

[7] Lohse, J.; Winteler, S.; Wulf-Schnabel, J. Collection targets for waste from electrical and electronic equipment (WEEE) the directorate general (DG XI) environment. Nuclear safety and civil protection of the Commission of the European Communities; 1998.

[8] Robinson, B. H. *Sci. Total. Environ.,* 2009, 183-191.

[9] Gaidajis, G.; Angelakoglou, K.; Aktsoglou, D. *J. Eng. Sci. Technol. Review*, 2010, 3, 193-199.

[10] Widmer, R.; Oswald-Krapf, H.; Sinha-Khetriwal, D.; Schnellmann, M.; Boni, H. *Environ. Impact Asses*., 2005, 25, 436-458.

[11] Electronics Industry Environmental Roadmap, Microelectronics and Computer Technology Corporation (MCC), Austin TX, 1996.

[12] Puckett, J.; Smith, T. Exporting harm: the high-tech trashing of Asia The Basel Action Network. Seattle7 Silicon Valley Toxics Coalition; 2002.

[13] Hageluken, C. Opportunities and Challenges to Recover Scarce and Valuable Metals from Electronic Devices, (2008) OECD-UNEP Conference on Resource Efficiency, Paris, Aprl 24, 2008.

[14] Dimitrakakis, E.; Janz, A.; Bilitewski, B.; Gidarakos, E. *J. Hazard. Mater.*,2009, 161, 913-919.

[15] OECD. Extended producer responsibility: a guidance manual for governments. Paris7 OECD; 2001.
[16] EU. Directive 2002/95/EC of the European Parliament and of the Council of 27 January 2003 on the restriction of the use of certain hazardous substances in electrical and electronic equipment (RoHS). Official Journal L037, 13/02/2003 p.19– 23; 2002b (*http://europa.eu.int/eur-lex/en/).*
[17] Wong, M. H.; Wu, S. C.; Deng, W. J.; Yu, X. Z.; Luo, Q.; Leung, A. O. W.; Wong,C. S. C.; Luksemburg, W. J.; Wong, A. S. *Environ. Pollut.* 2007, 149, 131-140.
[18] Schmidt, C.W. *Environ. Health Perspect.* 2006,114, A232-A235.
[19] Barba-Gutierrez, Y.; Adenso-Diaz, B.; Hopp, M. *Resour. Conser. Recycl.* 2008, 52, 481-495.
[20] UNEP. Basel convention on the control of transboundary movements of hazardous wastes and their disposal, United Nations Environment Programme/Secretariat of the Basel Convention; 1989. *http://www.basel.int/text/documents.html.*
[21] US Environmental Protection Agency website, *http://www.epa.gov/ecycling/docs/ fact11-07.pdf.*
[22] The World Bank, World Development Indicators (WDI), 2005.
[23] Environment Bureau Hong Kong SAR, Government, Consultation Document. Safe and Sustainable: *A New Producer Responsibility Scheme for Waste Electrical and Electronic Equipment*, 2009.
[24] Herrera, M.; Matuschek, G.; Kettrup, A.; *J. Anal. Appl. Pyrolysis*, 2003, 70, 35-42.
[25] Chen, L.; Wang, Y. -Z. *Polym. Adv. Technol.,* 2010, 21, 1-26.
[26] Schlummer, M.; Brandl, F.; Maurer, A.; van Eldik, R. *J. Chromatogr.*, A 2005, 1064, 39–51.
[27] Jr. Coombs, C.F. Printed Circuits Handbook 6th Ed., McGraw-Hill Professional, *Section,* 5.3, 2007.
[28] Jr. Coombs, C.F. Printed Circuits Handbook 6th Ed., McGraw-Hill Professional, *Section*, 12.4, 2007.
[29] Lassen, C.; Lokke, S. Brominated flame retardants substance flow analysis and assessment of alternatives, Danish Environmental Protection Agency, Copenhagen, Denmark, 1999.
[30] Weil, E.; Levchik, S. *J. Fire Sci.*,2004, 22, 25-40.
[31] ENIMCO, PCB Raw Materials, *http://www.enimco.com/materi als.html.*
[32] Grause, G.; Furusawa, M.; Okuwaki, A.; Yoshioka, T. *Chemosphere*, 2008, 71, 872-878.
[33] Abtew, M.; Selvaduray, G. " Mater. Sci. Eng. R: Report 2000, 27, 95-141.
[34] Prasad, L.C.; Xie, Y.; Mikula, A. *J. Non-Cryst. Solids,* 1999, 250, 1/- (doi: 10.1016/S0022-3093(99)00255-0).
[35] Suganuma, K.; Kim, S.-J.; Kim, K.-S. *JOM* 2009, 61, 64-71.
[36] Goosey, M.; Kellner, R. *Circuit World* 2003, 29, 33-37.
[37] Laner, D.; Rechberger, H. *Resour. Conser. Recycl.*, 2007, 52, 136–155.
[38] Nnorom, I. C.; Osibanjo, O. *J. Hazard. Mater.*, 2009, 161, 183-188.
[39] Chancerel, P.; Meskers, C.E.M.; Hagelüken, C.; Rotter, V.S. *J. Ind. Ecol.*, 2009, 13, 791-810.
[40] Veenstra, A.; Wang, C.; Fan, W.; Ru, Y. *Int. J. Adv. Manuf. Technol.*, 2010, 47, 449–459.

[41] Dwivedy, M.; Mittal, R. K. *Waste Manage.* 2010, 30, 483–491.
[42] Osibanjo, O.; Nnorom, I.C. *Environ. Impact Asses.*, 2008, 28, 198-213.
[43] Liu, X.; Tanaka, M.; Matsui, Y. *Waste Manage. Res.*, 2006, 24, 434-445.
[44] Liu, Q.; Li, K. Q.; Zhao, H.; Li, G.; Fan, F. Y. *Environ. Sci. Pollut. Res.*, 2009, 16, 248-249.
[45] Yu, X. Z.; Gao, Y.; Wu, S. C.; Zhang, H. B.; Cheung, K. C.; Wong, M. H. *Chemosphere,* 2006, 65, 1500-1509.
[46] Qu, W.; Bi, X.; Sheng, G.; Lu, S.; Fu, J. Yuen, J.; Li, L. Envrion. *Inter.*, 2007, 33, 1029-1034.
[47] Xing, G. H.; Chan, J. K. Y.; Leung, A. O. W.; Wu, S. C.; Wong, M. H. *Envrion. Inter.*, 2009, 35, 76-82.
[48] Brebu, M.; Bhaskar, T.; Murai, K.; Muto, A.; Sakata, Y.; Uddin, M. A. *Chemosphere* 2004, 56, 433-440.
[49] Mitan, N. M. M.; Bhaskar, T.; Hall, W. J.; Muto, A.; Williams, P. T.; Sakata, Y. *Chemosphere,* 2008, 72, 1073-1079.
[50] Leung, A. O. W.; Duzgoren-Aydin, N. S.; Cheung, K. C.; Wong, M. H. *Environ. Sci. Technol.*, 2008, 42, 2674-2680.
[51] Duarte, A. T.; Dessuy, M. B.; Silva, M. M.; Vale, M. G. R.; Welz, B. *Microchem. J.*, 2010, 96, 102-107.
[52] Deng, W.J.; Louie, P.K.K.; Liu, W.K.; Bi, X.H.; Fu, J.M.; Wong, M.H. *Atmos. Environ.*, 2006, 40, 6945–6955.
[53] Li, Y.; M. ASCE, P. E.; Richardson, J. B.; Walker, A. K.; Yuan, P. -C. *J. Environ. Eng.*, 2006, 497-504.
[54] Wang, D.; Cai, Z.; Jiang, G.; Leung, A.; Wong, M. H.; Wong, W. K. *Chemosphere,* 2005, 60, 810-816.
[55] Cai, Z.; Jiang, G. *Talanta*, 2006, 70, 88-90.
[56] Leung, A. O. W.; Luksemburg, W. J.; Wong, A. S.; Wong, M. H. *Environ. Sci. Technol.*, 2007, 41, 2730-2737.
[57] Sepulveda, A.; Schluep, M.; Renaud, F. G.; Streicher, M.; Kuehr, R.; Hageluken, C.; Gerecke, A. C. *Environ. Impact Asses.*, 2010, 30, 28–41.
[58] Leung, A.; Cai, Z. W.; Wong, M. H. *J. Mater. Cycl. Waste Manage.* 2006, 8, 21-33.
[59] Zhao, G.; Xu, Y.; Han, G.; Ling, B. *Environ. Geo. Health*, 2006, 28, 341–351.
[60] Huo, X.; Peng, L.; Xu, X.; Zheng, L.; Qiu, B.; Qi, Z.; Zhang, B.; Han, D.; Piao, Z. *Environ. Health Perspect.*, 2007, 115, 1113-1117.
[61] Li, Y.; Xu, X.; Liu, J.; Wu, K.; Gu, C.; Shao, G.; Chen, S.; Chen, G.; Huo, X. *Sci. Total Environ.*, 2008, 403, 99-104.
[62] Guo, Y.; Huo, X.; Li, Y.; Wu, K.; Liu, J.; Huang, J.; Zheng, G.; Xiao, Q.; Yang, H.; Wang, Y.; Chen, A.; Xu, X. Sci. *Total Environ.*, 2010, 408, 3113-3117.
[63] Liu, Q.; Cao, J.; Li, K. Q.; Miao, X. H.; Li, G.; Fan, F. Y.; Zhao, Y. C. *Environ. Sci. Pollut. Res.*, 2009, 16, 329-338.
[64] Wang, H.; Zhang, Y.; Liu, Q.; Wang, F.; Nie, J.; Qian, Y. *Inter. J. Hyg. Environ. Health*, 2010, 213, 369-380.
[65] The Encyclopedia of Earth, Computer Recycling, *http://www.eoearth.org/article/Computer_recycling*.
[66] Li, J.; Lu, H.; Guo, J.; Xu, Z.; Zhou, Y. *Environ. Sci. Technol.* 2007, 41, 1995-2000.

[67] Long, L.; Sun, S.; Zhong, S.; Dai, W.; Liu, J.; Song, W., *J. Hazard. Mater*. 2010, 177, 626-632.

[68] Szczygiel, Z.; Lara, C.; Escobedo, S.; Mendoza, O. *JOM* 1998, 50, 55-59.

[69] Oishi, T.; Koyama, K.; Alam, S.; Tanaka, M.; Lee, *J.-C. Hydrometallurgy,* 2007, 89, 82-88.

[70] Mecucci, A.; Scott, K. *J. Chem. Technol. Biotechnol*., 2002, 77, 449-457.

[71] Veit, H.M.; Diehl, T.R.; Salami, A.P.; Rodrigues, J.S.; Bernardes, A.M.; Tenório, *J.A.S. Waste Manage*. 2005, 25, 67-74.

[72] Eswaraiah, C.; Kavitha, T.; Vidyasagar, S.; Narayanan, S.S. *Chem. Eng. Process.: Process Intensification*, 2008, 47, 565-576.

[73] Cui, J.; Forssberg, E. *J. Hazard. Mater*., 2003, 99, 243-263.

[74] Veit, H.M.; Pereira, C.D.C.; Bernardes, *A.M. JOM* 2002, 54, 45-47.

[75] Huang, K.; Guo, J.; Xu, Z. *J. Hazard. Mater*., 2009, 164, 399-408.

[76] Zhang, S.; Forssberg, E. *Resour., Conserv. Recycl*., 1997, 21, 247-269.

[77] Moltó, J.; Font, R.; Gálvez, A.; Conesa, J.A. *J. Anal. Appl. Pyrolysis*, 2009, 84, 68-78.

[78] Blazsó, M.; Czégény, Z.; Csoma, C. *J. Anal. Appl. Pyrolysis*, 2002, 64, 249-261.

[79] Barontini, F.; Marsanich, K.; Petarca, L.; Cozzani, V. *Ind. Eng. Chem. Res*.,2005, 44, 4186-4199.

[80] Jie, G.; Ying-Shun, L.; Mai-Xi, L. *J. Anal. Appl. Pyrolysis,* 2008, 83, 185-189.

[81] Hall, W.J.; Williams, P.T. *Resour. Conserv. Recycl*., 2007, 51, 691–709.

[82] Gloe, K.; Mühl, P.; Knothe, M. *Hydrometallurgy* 1990, 25, 99–110.

[83] Das, A.; Vidyadhar, A.; Mehrotra, S.P. *Resour. Conserv. Recycl*., 2009, 53, 464-469.

[84] Xiu, F. –R.; Zhang, F. –S. *J. Hazard. Mater*., 2010, 178, 1-3, 628-634.

[85] Yoo, J. –M.; Jeong, J.; Yoo, K.; Lee, J.-C.; Kim, W. *Waste Manage*., 2009, 29, 1132-1137.

[86] Vetri Murugan, R.; Bharat, S.; Deshpande, A. P.; Varughese, S.; Haridoss, P. *Powder Technol*., 2008, 183, 169-176.

[87] Koyanaka, S.; Endo, S.; Ohya, H. *Adv. Powder Technol*., 2006, 17, 113-126.

[88] Xiu, F. -R.; Zhang, F. -S. *J. Hazard. Mater*., 2009, 172, 1458-1463.

[89] Li, J.; Shrivastava, P.; Gao, Z.; Zhang, H. -C. *IEEE Trans. Electron. Packaging Manuf*., 2004, 27, 33-42.

[90] Gibson, R.W.; Fray, D.J.; Sunderland, J. G.; Dalrymple, I. M. *Electrochem. Soc*. 2003, 346–354.

[91] Feldmann, K.; Scheller, H. in Concept— Conf. Clean Electronic Products and Technol.,1995, IEE Conf. Publication no. 415, 186–190.

[92] Zhou, Y.; Qiu, K. *J. Hazard. Mater*., 2010, 175, 823-828.

[93] Guo, J.; Rao, Q.; Xu, Z. *J. Hazard. Mater*., 2008, 153, 728–734.

[94] Hong, S.G.; Su, S. H. J. *Environ. Sci. Health,* 1996, A31, 1345–1359.

[95] Zheng, Y.; Shen, Z.; Cai, C.; Ma, S.; Xing, Y. *J. Hazard. Mater,*. 2009, 163, 600–606.

[96] Zheng, Y.; Shen, Z.; Cai, C.; Ma, S.; Xing, Y. *Mater. Des*., 2009, 30, 958–963.

[97] Dang, W.; Kubouchi, M.; Sembokuya H.; Tsuda, K. *Polymer,* 2005, 46, 1905–1912.

[98] Zheng, Y.; Shen, Z.; Ma, S.; Cai, C.; Zhao, X.; Xing, Y. *J. Hazard. Mater,*. 2009, 170, 978-982.

[99] Owens, C.V.; Lambright, C.; Bobseine, K.; Ryan, B.; Gray, L.E.; Gullett, B.K.; Wilson, V.S. *Environ. Sci. Technol*. 2007, 41, 8506–8511.

[100]Huang, Y.; Takaoka, M.; Takeda N.; Oshita, K. *Environ. Sci. Technol.*, 2007, 41, 257–262.

[101]Kang, H.Y.; Schoenung, J.M. *Resour. Conserv. Recycl.*, 2005, 45, 368–400.

[102]Barontini, F.; Cozzani, V. *J. Anal. Appl. Pyrolysis,* 2006, 77, 41-55.

[103]Chiang, H. -L.; Lin, K. -H.; Lai, M. -H.; Chen, T. -C.; Ma, S. -Y. *J. Hazard. Mater.*, 2007, 149, 151-159.

[104]Guo, J.; Guo, J.; Xu, Z. *J. Hazard. Mater.*, 2009, 168, 567-590.

[105]Vasile, C.; Brebu, M.A.; Totolin, M.; Yanik, J.; Karayildirim, T.; Darie, H. *Energy Fuels,* 2008, 22, 1658–1665.

[106]Van den Berg, M.; Birnbaum, L. S.; Denison, M.; et al. *Toxicol. Sci.*, 2006, 93, 223-241.

[107]Schecter, A.; Birnbaum, L.; Ryan, J. J.; Constable, J.D. *Environ. Res.*, 2006, 101, 419-428.

[108]Kulkarni, P. S.; Crespo, J. G.; Afonso, C. A. M. *Environ. International,* 2008, 34, 139-153.

[109]Sato, Y.; Kodera, Y.; Kamo, T. *Energy Fuel*, 1999, 13, 364-368.

[110]Kawai, N.; Tsujita, K.; Kamo, T.; Sato, Y. Chemical recovery of bisphenol-a from polycarbonate resin and waste. In: Proceedings of ISFR'2002 the 2nd international symposium on feedstock recycling of plastics and other innovative recycling techniques, September 8–11 2002, Ostend, Belgium, CD-A48, 2002.

[111]Van Schindel, P. P. A. J.; van Kasteres, J. M. N. Design and optimisation of a separation, purification and upgrading process for polymers from electronic and electric equipment. In: Proceedings of ISFR'2002 the 2nd International symposium on feedstock recycling of plastics and other innovative recycling techniques, September 8–11 2002, Ostend, Belgium, CD-A33, 2002.

[112]Agnes F. The Watech process for treatment of plastic waste with brominated flame-retardants. In: Proceedings of ISFR'2002 the 2nd international symposium on feedstock recycling of plastics and other innovative recycling techniques, September 8–11 2002, Ostend, Belgium, CD-A86, 2002.

[113]Bockhorn, H.; Hornung, A.; Hornung, U.; Jakobstroer, P.; Kraus, M. *J. Anal. Appl. Pyrolysis,* 1999, 49, 97-106.

[114]Boerrigter, H.; André, B.; Oudhuis, J.; Tange, L. Bromine recovery from the plastics fraction of waste of electrical and electronic equipment (WEEE) with staged gasification, Paper R'02 WEE in Pyromaat, Netherland, 2002.

[115]Bhaskar, T.; Kaneko, J.; Muto, A.; Sakata, Y.; Jakab, E.; Matsui, T. et al. *J. Anal. Appl. Pyrolysis,* 2004, 71, 765–777.

[116]Bhaskar, T.; Kaneko, J.; Muto, A.; Sakata, Y.; Jakab, E.; Matsui, T. et al. *J. Anal. Appl. Pyrolysis,* 2004, 72, 27–33.

[117]Sakata, Y.; Bhaskar, T.; Uddin, M.D.A.; Muto, A.; Matsui, T. *J. Mater. Cycles Waste Manage.*, 2003, 5, 113–124.

[118]Bhaskar, T.; Murata, K.; Matsui, T.; Brebu, M.A.; Uddin, M.D.A.; Muto, A. et al. *J. Anal. Appl. Pyrolysis,* 2003, 70, 369–381.

[119]Vasile, C.; Brebu, M. A.; Karayildirim, T.; Yanik, J.; Darie, H. *Fuel,* 2007, 86, 477-485.

[120]Blazso, M.; Czegeny, Z.; Csoma, C. *J. Anal. Appl. Pyrolysis,* 2002, 64, 249-261.

[121]Hornung, A.; Donner, S.; Balabanovich, A.; Seifert, H. *J. Cleaner Production*, 2005, 13, 525-530.

[122]Guan, G.; Zhou, W.; Chen, L.; Chen, Y.; Cai, M. *CIESC J.* 2009, 61, 216-222.

[123]Peng, S.-H.; Chen, L.-Q.; Hu, Y.-L.; Yao, X.-Q. *Modern Chem. Ind.*, 2009, 29, 44-47.

[124]Lai, Y.-C.; Lee, W.-J.; Li, H.-W.; Wang, L.-C.; Chang-Chien, G.-P. *Environ. Sci. Technol.*, 2007, 41, 957-962.

[125]Weifang Qiangyuan Chemical Industry Co., Ltd., Shandong, China (Mainland) *http://www. alibaba.com/product-gs/286404150/ Calcium_Bromide_ 14_2ppg.html.*

[126]Sino Rich Chemical Co., Ltd., Shandong, China (Mainland), *http://www. alibaba.com/product-gs/317151870/Calcium_Bromide.html.*

[127]Zhejiang Jun Xiang Mining Co., Ltd., Zhejiang, China (Mainland), *http://www. alibaba.com/product-gs/309327805/calcium_ carbonate.html.*

[128]Tianjin Sbarter ImpandExp Co., Ltd., Tianjin, China (Mainland), *http://www. alibaba.com/product-gs/334277853/Calcium_ Carbonate.html.*

[129]de Marco, I.; Caballero, B.M.; Chomon, M.J.; Laresgoiti, M.F.; Torres, A.; Fernandez, G.; Arnaiz, S. *J. Anal. Appl. Pyrolysis.* 2008, 82, 179-183.

[130]Norris, D. R. Development of the Pyrolysis Process for Recycling of SMC, *Eagle-Picher Plastics División* pp 277, 1992.

[131]SMC Automotive Alliance U.S.A., Proceeding of the 46th Annual Conference Composites Institute, The Society of the Plastics Industry Inc., Washington, DC, p. 18, 1991.

[132]*http://www.essom.com/ backend/data-file/engineer/engin21_1.pdf.*

[133]Cui, J.; Zhang, L. *J. Hazard. Mater..* 2008, 158, 228-256.

[134]Castro, L.A.; Martins, A.H. Braz. *J. Chem. Eng..* 2009, 26, 649-657.

[135]Liu, R.; Shieh, R. S.; Yeh, R. Y. L.; Lin, C. H. *Waste Manage.*, 2009, 29, 2842-2845.

[136]Li, J.; Lu, H.; Xu, Z.; Zhou, Y. *J. Hazard. Mater.*, 2008, 151, 52-57.

[137]Li, J.; Xu, Z. *Environ. Sci. Technol.*, 2010, 44, 1418-1423.

[138]Veit, H. M.; Bernardes, A. M.; Ferreira, J. Z.; Tenorio, J. A. S.; Malfatti, C. D. F. *J. Hazard. Mater.*, 2006, 137, 1704-1709.

[139]Ivanuş, R.C. *Metalurgia International*, 2010, 15, 10-14.

[140]Havlik, T.; Orac, D.; Petranikova, M.; Miskufova, A.; Kukurugya, F.; Takacova, Z. *J. Hazard. Mater.*, 2010, 183, 866-873.

[141]National Physical Laboratory, Chapter 3: Chemistry, 3.11.4 Calorific values of solid, liquid and gaseous fuels, http://www.kayelaby.npl.co.uk/chemistry/3_11/3_11_4.html.

[142]Quan, C.; Li, A.; Gao, N. *J. Hazard. Mater.*, 2010, 179, 911-917.

[143]Veldbuizen, H.; Sippel, B. *Ind. Environ.*, 1994, 17, 7-11.

[144]Theo, L. Integrated recycling of non-ferrous metals at Boliden Ltd. Ronnskar smelter, *IEEE International Symposium on Electronics and the Environment*, 42–47.

[145]Dunn, J.; Wendell, E.; Carda, D.D. et al. Chlorination process for recovering gold values from gold alloys, US Patent, US5004500 (1991).

[146]Aleksandrovich, S.; Nicolaevich, E.; Ivanovich, E. Method of processing of products based on ahalcogenides of base metals containing metals of platinum group and gold, *Russian Patent*, RU2112064, C22B 11/02 (1998).

[147]Hageluken, C. Recycling of electronic scrap at umicore's integrated metals smelter and refinery, *World of Metallurgy – ERZMETALL*, 2006, 59, 152–161.

[148]Wai, C.M.; Wang, J.S. *Abstracts of Papers of the American Chemical Society,* 229, U930-U931, Mar 13, 2005.

[149]Baba, H. *Conserv. Recycl.*, 1987, 10, 247–252.

[150]Macaskie, L.E.; Creamer, N.J.; Essa, A.M.M. et al. *Biotechnol. Bioeng*., 2007, 96, 631-639.

[151]J. Vinals, E. Juan, A. Roca et al., The metalozon [registered trademark] process for the recovery and recycling of precious metals, REWAS'04, 2004– Global Symposium on Recycling, *Waste Treatment and Clean Technology,* 1859–1869.

[152]Paretsky, V.M.; Antipov, N.I.; Tarasov, A.V. Hydrometallurgical method for treating special alloys, jewelry, electronic and electrotechnical scrap, *TMS Annual Meeting*, 2004, 713–721.

[153]Andrews, D.; Raychaudhuri, A.; Frias, C. *J. Power Sources,* 2000, 88, 124–129.

[154]Faramarzi, M. A.; Stagars, M.; Pensini, E.; Krebs, W. J. *Biotechnol.*, 2004, 113, 321-326.

[155]Brandl, H.; Bosshard, R.; Wegmann, M. *Hydrometallurgy* 2001, 59, 319-326.

[156]Bosshard, P.P.; Bachofen, R.; Brandl, H. *Environ. Sci. Technol.* 1996, 30, 3066-3070.

[157]Brombacher, C.; Bachofen, R.; Brandl, H. *Appl. Environ. Microbiol.* 1998, 64, 1237-1241.

[158]Suzuki, *I. Biotechnol. Adv*., 2001, 19, 119-132.

[159]Brandl, H.; Lehmann, S.; Faramarzi, M.A.; Martinelli, D. *Hydrometallurgy* 2008, 94, 14-17.

[160]Tributsch, H. *Hydrometallurgy,* 2001, 59, 177-185.

[161]Bosshard, R. Metallrückgewinnung Aus Elektronikschrott Mit Hilfe Von. Bakterien Diploma thesis, ETH Zürich, Switzerland, 1998.

[162]Reuter, M.A.; Van Schaik, A.; Ignatenko, O.; De Haan, G. *J. Miner. Eng*. 2006, 19, 433-449.

[163]Quinet, P.; Proost, J.; Van Lierde, A. *Miner. Metall. Process*., 2005, 22, 17-22.

[164]Park, Y.J.; Fray, D.J. *J. Hazard. Mater*., 2009, 164, 1152-1158.

[165]Yan, F. L. Metals Recovery from Scrap IC Board, Energy Resource Laboratory (1998).

[166]http://www. metalprices.com/ FreeSite/metals /cu/cu.asp.

[167]Times Nano http://www.timesnano.com/info/price.html.

[168]Helix http://www.helixmaterial. com/Ordering.html.

In: E-Waste: Management, Types and Challenges
Editors: Yuan Chun Li and Banci Lian Wang
ISBN 978-1-61942-217-9

Chapter 3

BRIDGING THE DIGITAL DIVIDE AND CREATING AN ICT DUMP: AN OVERVIEW OF THE UNSUSTAINABILITY OF EXPORTING USED AND END-OF-LIFE ICT-WARES TO DEVELOPING COUNTRIES

Nnorom Innocent Chidi
Department of Industrial Chemistry, Abia State University,
Uturu, Abia State, Nigeria

ABSTRACT

Digital divide refers to the gap between those who have access to technology and those who do not. Increasing quantities of used electronic products are exported to some developing countries annually; a greater proportion being non-functional and un-usable. Consequently, large quantities of electronic waste (e-waste) are being managed in developing countries using inappropriate management methods including: disposal at open dumps, open burning and the use of crude extraction techniques in recovering valuables. Reasons given for the trans-boundary movement of e-waste appears plausible and laudable - it will benefit those who are too poor to afford brand new equipment thereby *bridging the digital divide*, result in environmental gains as cheaper labour in the developing countries can make repair of old equipment feasible, giving it longer life, and even recycle e-waste. The dumping of e-waste in developing countries is unsustainable, environmentally unjust and undermines the health of the ecosystem. This practice exposes developing countries to an array of highly toxic components of e-waste. The digital divide can be bridged in sustainable ways that provide digital opportunity for the poor without worsening their condition. This can be achieved by assisting developing countries with 'tested' and "functional" "used electronic devices" and not 'electronic scrap' to be recycled. There is therefore the need for a global policy initiative and legislation in dealing with this emerging global crisis.

Keywords: Environmental pollution; electronic waste transboundary shipments; digital divide, waste exports

1. Introduction

The concept of the "digital divide" expresses the gap in access to information resources in some countries compared to those with state-of-the-art networks: telephone, radio, TV, Internet, satellite, in short, anything that can be classed as information communication technologies (ICTs) (Ya'u, 2002; Marine and Blanchard, 2004). In other words, it is the unequal access by some members of society to ICT and the unequal acquisition of related skills.

The term is closely related to the knowledge divide as the lack of technology causes lack of useful information and knowledge. Consumer electrical and electronic appliances have been instrumental to the revolution witnessed in diverse areas of human endeavours including communication, entertainment, transportation, education, and health care delivery (Osibanjo and Nnorom, 2007).

There appears to be no signs of this revolution slowing down soon. In fact it is accepted that technological innovations will continue to be a cornerstone of social progress and advanced electronic products will continue to detect the pace and direction. It has been acknowledged that ICT have the potential to support sustainable development (Hilty et al, 2006). The effective application of ICT has the potential to benefit all aspects of life - from agriculture to environmental protection, from the management of natural resources, to disaster prevention, and in supporting sustainable production and consumption patterns (Lall and Garai, 2005).

While there are many factors contributing to the digital divide, the high price of information technology hardware is no doubt a significant one (Williams, 2003). Due to lack of financial resources to most people in developing countries, much of the growth in the ICT sector rely on second-hand equipment imported from rich, developed countries. Presently, large quantities of used electronics are exported to developing countries to help bridge the digital divide.

Majority of the exported goods are non-functional (e-waste) and are posing serious management challenges in the developing countries (Nnorom and Osibanjo, 2008). E-waste has evolved into a global environmental problem and the main challenges is that the infrastructure to manage it properly is still not well developed. The recycling of some components of e-waste such as the cathode ray tube is technically problematic and not feasible economically. In most developing countries recycling e-waste requires huge investments.

This paper reviews the challenges faced by developing countries from the widening digital divide and the benefits accruable from assisting them with ICT-wares. It also provides an overview of the origin and philosophy of trans-boundary movement of e-waste into developing countries. Subsequent sections outline the environmental and health consequences of the adoption of unsound management practices in e-waste management in the e-waste receiving countries.

Issues in finding a global solution to the e-waste crisis were also presented. This is critical to adopting a sustainable globally acceptable management options for e-waste.

2. The ICT Digital Divide

2.1. ICT and Socioeconomic Development of Developing Countries

Access to ICT has been identified as an indicator of a country's economic and social situation. Digital divide would not have attracted so much attention were it not for its impact on development within a global economy, increasingly based on the exchange of information and knowledge (Marine and Blanchard, 2004). In appreciating the relevance of ICT and the importance of bridging the digital divide, the UN Secretary-General Ban Ki-Moon encouraged member states to bridge the digital divide by turning " the digital divide into digital opportunity" (Jhin, 2007).

Obijiofor (2009) reviewed theoretical and practical issues on the relationships that exist between ICT implementation and Africa's socioeconomic development and acknowledged that access to new technologies (computers, mobile phones, Internet, telecentres, etc.) remains the biggest obstacle to wider diffusion and use of new technologies in Africa. He observed that:

> "it is access to the new technologies that determines whether or not people are able to use new technologies. Education is important but not a prerequisite because educationally illiterate people are known to be using mobile phones to communicate in their local languages. So, access is a priority. And unless access to new technologies is treated as a priority problem, efforts by governments, community groups and aid agencies to implement policies on diffusion of new technologies would continue to encounter obstacles" (Obijiofor, 2009).

Most developing countries are struggling to bridge the divide with Africa at the bottom of the Internet ladder (Figure 1). The effects of the digital divide on countries and individuals includes unfair competition, higher costs of access to ICT as well as the cost of maintaining the divide and unfair competition in the global playing field (Table 1). Having observed the role of ICT in advancing development, most countries and individuals queued up eager to access the benefits of ICT

Table 1. Effects of the digital divide on countries and individuals

Issue	Notes
Unfair competition	Disadvantaged countries are unable to compete with their counterparts in the developed countries due to lack of access to information and knowledge (e.g. access to funding for research, scholarships, etc).
Cost of access to ICT	Worsening/deepening poverty as the digitally disadvantaged spends more to access the technology, (higher costs of access to the Internet and phone calls).
Unfair playing field	The digital divide reduces the possibility of a global "level field" in trade, investment and relations, job opportunities etc.
Maintaining the divide	Higher costs of "maintaining" the digital divide or even attempting to bridge it has negative effects on economic stability.

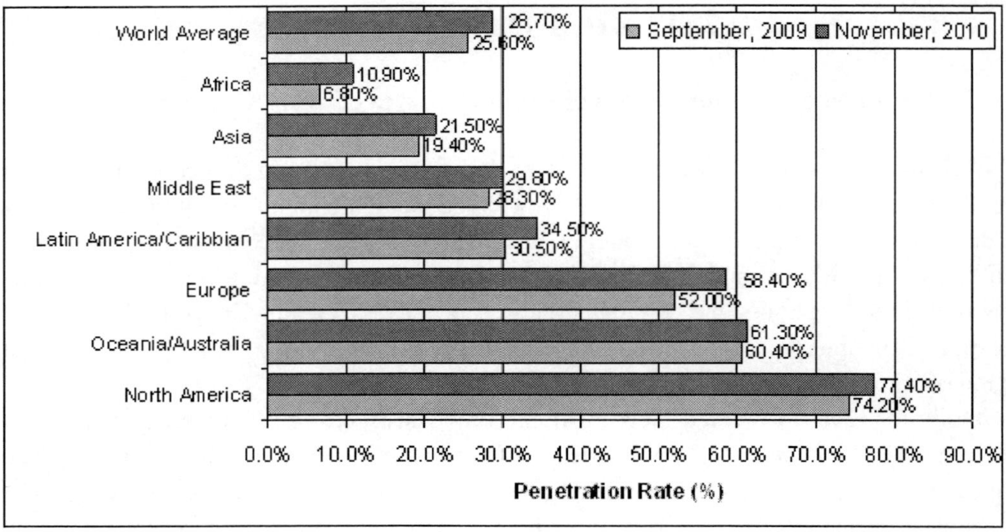

Figure 1. Internet penetration.

2.2. The Digital Divide

Digital divide is defined as "*the unequal access to ICT to people relative to their geographical location, living standard, level of education, and sex*" (Ya'u, 2002; Marine and Blanchard, 2004). It refers to the differences associated with access to and use of applications of ICTs among individuals of a given country, or between different countries or regions (Monge and Chacón, 2002). It also refers to the gap that exists in the opportunities to access advanced information and communication technologies between geographic areas or by individuals at different socio-economic levels. In fact, Monge and Chacón (2002) observed that the concept of a digital divide is closely related to income, gender, educational level, place of residence, age, language, physical disabilities and other socioeconomic dimensions. Indices of the digital divide include PC penetration, Internet usage, bandwidth consumption, content creation and online representation.

There is a large digital divide in most developing countries, and only a small number of wealthy people have access to certain ICT-wares especially branded new equipment. Consequently, most of the low-income earners rely on used ICT-wares (and cloned brands) as branded new equipment are relatively expensive for ordinary people. Because most of the population do not have access to ICT-ware or have the knowledge, effective use of ICT in key sectors such as governance and commerce, education and healthcare are lacking in these countries. For instance, electronic commerce is not flourishing in most developing countries, because credit cards are not generally accepted or used. Consequently, the World Summit on the Information Society sought to provide an overarching framework for harnessing the potential of ICTs in the pursuit of development goals, including that of ensuring sustainable development.

2.3. Global and Regional Efforts at Bridging the Digital Divide

The realization of the importance of ICTs in economic development led the UN Commission on Science and Technology for Development (UNCSTD) to devote the years 1995-1997 to the study of the linkage between ICTs and development. One of the practical results of that effort was the placing of the digital divide on global development agenda. Since then, there has been a consensus internationally that there is the need to bridge the digital divide. Multiple initiatives have been, and continue to be, implemented across the world to address the major obstacles to bridging the digital divide. The digital divide was one of the central topics in the Group of Eight Developed Countries (G8) Summit talks held in Kyushu and Okinawa in July 2000. The G8 governments proposed, "ICT (IT) as one of the most potent forces in shaping the twenty-first century." During the Summit, the G8 established the 'Okinawa Charter on Global Information Society', an initiative to bridge the 'digital divide', and improve access to ICT in the world's poorer countries.

The United Nations organized the World Summit on the Information Society (WSIS) in December 2003. The first WSIS summit adopted a declaration and an action plan to make ICT accessible to everyone in the interest of development. The declaration emphasized ICTs role as a development tool and enshrined the objective of universal access to ICT in the wider context of the United Nations Millennium Development Goals, MDGs (Maine and Blanchard, 2004). The United Nations Global Alliance for Information and Communication Technologies and Development was also established to achieve a similar objective. The Mission of the Global Alliance for ICT and Development (GAID) is to contribute to transforming the spirit and vision of WSIS into action and promoting the use of ICT for the achievement of the internationally agreed development goals, including the MDGs (Jhin, 2007).

In Africa, the United Nations Economic Commission for Africa (UNECA) initiative of May 1996 resulted in the resolution titled "Building Africa's Information Highway" which committed member countries to the building of national ICT. The Report of the committee is called "Africa Information Society Initiative, (AISI)" (Ya'u, 2002).

The export and donation of cheap, used and refurbished electronics have been recognized as important means for bridging the digital divide between developed and developing countries. In line with this, other private sector driven programmes have been launched with intention of bridging the digital divide. Some of these programmes are aimed at assisting educational institutions in the developing countries with low cost new computers, second hand and refurbished computers. These programmes include;

- 'World Computer Exchange'
- 'Computer Aid',
- 'Close the Gap' and
- 'One Laptop Per Child' (OLPC).

These programmes have been in the forefront of assisting the developing countries with ICT wares.

2.4. Bridging the Divide or Widening the Gap?

Many developing countries have poor communications infrastructure, which prevents connectivity, particularly in the rural areas. The issue of infrastructure is a major issue, as this requires global, long-term investment, and support from a variety of sources. The developing countries lose unquantifiable wealth annually to the Digital Divide due to the high cost of communication and Internet services. Consequently, these countries rely on the developed countries for access and connectivity with the attendant 'security risks'.

The migration of the few educated and ICT-competent persons from the developing to developed countries (brain drain) and the unwillingness of those trained in universities in developed countries to return home contribute to the widening divide. In the decade 2001-2010, Africa had the highest growth in Internet access (Figure 2).

This not withstanding, the continent is still at the bottom of the ladder with a penetration rate about a third of the world average (see Figure 1). Despite the initiatives to bridge the divide, the gap appears to be increasing rather than decreasing.

Despite the boom in the availability of access to communication resources since the beginning of the 1990s (from 700 million to nearly 2.5 billion telephone lines, and from scarcely a million to over a billion Internet users), the divide in access to ICT has been deepening. In fact, most authors are of the opinion that the differences in the usage of communication resources between countries and regions have been intensifying (Maine and Blanchard, 2004; Ya'u, 2002).

In the developing countries, majority of users prefer used electronics because they are cheaper, and are perceived to be of high quality and of reliable brand, having been manufactured and used in the developed countries. Consequently, those needing to cut costs are motivated by the lower prices of used goods.

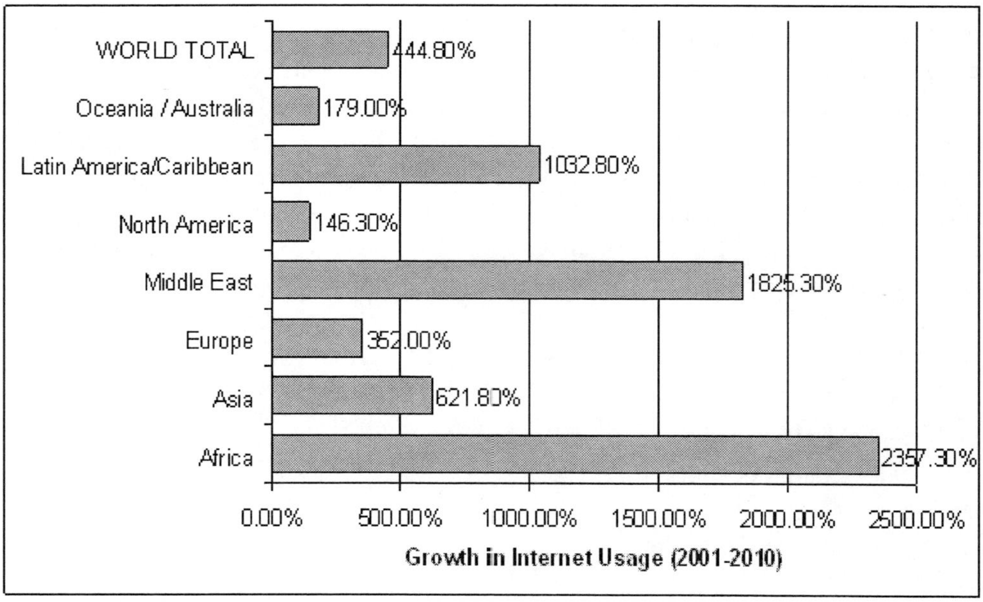

Figure 2. Growth in Internet Usage around the world.

The extension of the useful life of an electronic device should be a priority in the environmental management of electronics. Williams (2003) observed that one important and practical way to do this is through encouraging the market for used electronics as most users in the developed countries normally dispose their electronics before they become dysfunctional. Some users in developed countries discard old equipment to create space for new ones with better features and specifications, and not necessarily because the equipment are dysfunctional.

2.5. Benefits of Bridging the Digital Divide

Trade in used electronics products maybe necessary in closing the digital divide. It would provide the opportunity for those who may not be able to afford new devices especially low-income students and families to obtain low-cost working electronic devices often after repairs. This is because not all users require high performance, most of the popular applications of computers such as e-mail, and web browsing work well on older machines. Consequently, the waste hierarchy of "reduce, reuse, and recycle" changes to "reduce, reuse, repair, and recycle.

Progress in information revolution and ICT diffusion across the world has been of benefit to the persons who would not have had access especially in the developing countries. Most developing countries have shown serious concern at bridging the digital divide as well as instituted policies and programmes aimed at reducing the divide, especially focusing on increasing network infrastructure provision. In fact, the rapid and widespread diffusion of the information revolution in the developing countries have witnessed an enthusiastic embrace by so many especially in Africa, due to the failure of the old telecom systems to create a reliable, open and widespread system for transmitting information.

Table 2. Benefits of bridging the digital divide

Issue	Notes
Cost	The new ICTs is less expensive and more powerful as they bring greater access to information and knowledge;
Governance	The new ICT have been supporting e-governance around the globe. Increasing use of ICTs to provide information on election, voters' registration and candidates in national elections. For example SMS (short text messages) are sent to potential voters as part of the campaign strategy;
Business	Drastic revolutions in the way people in developing countries communicate, and conduct business. ICTs allow businesses to make more informed and strategic decisions as well as electronic commerce (e-commerce).
Education	ICT is opening educational opportunities for many including distance education. It also plays an important role as a tool in public education, and in pressing for changes in government policies;
Access to healthcare	Cheaper and better telecommunication services are also carrying telemedicine. ICT have assisted in rescue operations and in providing information to doctors in remote areas on life-saving operations/surgeries (Wilson and Wong, 2003).

The benefits of bridging the divide are enormous and this impacts positively on socio-economic life of the people (Table 2). The developing countries need ICT wares in technology transfer; technical assistance; capacity building; early warning systems; information or resource management systems; monitoring and assessment mechanisms; communication; education; awareness-raising; promoting and facilitating the exchange of information on best practices; knowledge sharing; dissemination of scientific or technical knowledge; and the wider application of techniques and assessment methodologies, etc (Lall and Garai, 2005).

In fact, many in the developing countries now use ICT wares especially the mobile phone and personal computer on a daily basis to interact with individuals and access information from around the globe.

ICT diffusion stands to assist developing countries in achieving the improvements that are so desperately needed, especially in the areas of economic and social changes, education, and technological advancement. However, if ICT diffusion promises improvements, it also poses substantial risks especially with respect to environmental pollution at the end-of-life (EoL) of these ICT devices if not appropriately handled. The performance improvements brought by ICT has resulted in increased consumption of ICT products and services. This has numerous environmental implications on different levels – at the material extraction level, manufacturing, and end-of-life phase.

3. Transboundary Movements of Used Electronics

3.1. Philosophy of Used Electronics Exports

Several reasons have been used to support the export of used electronics and electronic scrap from developed to developing countries. Some of the philosophies used to back the trade are (BAN, 2002; 2005):

- It will benefit those who are too poor to afford brand new equipment, thereby providing those more needy with a way to cross the digital-divide,
- it adds extra life to the product
- It will lead to environmental gains/protection as cheaper labour in the developing countries can make repair and re-use of the old equipment feasible, giving it a longer life, and,
- The re-use of second-hand devices will forestall the need for more products to be manufactured.
- Cheap labour will make repair, and recycle economically feasible while creating jobs

Others are of the view that the export of functional used EEE will benefits the receiving countries in:

- ♦ bridging the digital divide by providing cheap EEE to assist the operations of schools, small businesses, and government agencies while extending the life cycle of electronic products

- job creation in the ICT sector including repairs/refurbishing; import/export, clearing, local transportation/logistics
- Promoting sustainable development (reuse of components from those products, minimization of health and environmental risks).

However it is obvious that WEEE is increasingly been shipped out as second-hand goods from developed countries to the poor developing countries in order to avoid:

1. the high cost of landfilling e-waste, now classified as hazardous waste,
2. the potential environmental pollution from the landfilling (as well as recycling) and
3. the increasing un-profitability of e-waste recycling and /or inadequacy of recycling facilities (Kojima, 2005; Hageluken, 2006a; 2006b; BAN/SVTC, 2002; BAN, 2005; Greenpeace, 2005a).

3.2. Bridging the Digital Divide: Creating a Digital Dump

Used electronics are exported to poor underdeveloped countries on supposed benefits listed above, but unfortunately, exported equipment are rarely tested or confirmed 'functional'.

In addition, they are not distinguished from new EEE, or from scrap EEE meant for recycling. As a result, efforts at bridging the digital divide have degenerated into a 'pipeline' for e-waste disposal as unusable and scrap EEE are exported in the name of reuse. In fact, Williams, (2003) reviewed the e-waste issues and observed that:

> Imports of used or rather end-of-life (EoL) electronics to the developing countries is a double-edged sword: the positive potential also carries with it the risk of becoming a dumping ground for waste equipment from rich developed countries (Williams, 2003).

It has become obvious that a major route of disposal of e-waste in some developed countries appears to be via export to developing countries in the name of 'bridging the digital divide'.

The Basel Action Network (2005) observed that "too often, justifications of 'building bridges over the digital divide' are used as excuses to obscure and ignore the fact that these bridges double as toxic waste pipelines to some of the poorest communities and countries in the world. While supposedly closing the 'digital divide', the developed countries are rather opening a 'digital dump'" (BAN, 2005).

A considerable part of the e-waste exported from developed countries is illegal in contradiction to international (Basel Convention) or regional (EU waste shipment regulation; African Bamako Convention) legislation from either of the following perspectives:

- completely illegal shipment, where the rules applicable are not respected; or
- e-wastes that are shipped under a false pretence, e.g., e-waste reported to be post-consumer products for re-use, while they are clearly no longer fit for re-use.

The export of e-waste to Africa and Asia appears to be a preferred option to some developed countries rather than use the opportunity to enable their own national recycling infrastructure, switch to cleaner technologies and develop innovative design to prevent further toxics use. The developing countries are increasingly victimized by a disproportionate burden of the world's toxic cyber waste (BAN, 2005).

4. THE INCREASING GLOBAL ENVIRONMENTAL BURDEN OF E-WASTE

4.1. Environment Impact of Rapid ICT Diffusion

The relationship between the development, diffusion, use of ICTs and the broader social goal of sustainability is not well understood. Berkhout and Hertin (2004) examined available literature on the environmental impact of ICTs and the Internet, and identified three main types of effects (Table 3). The study concluded that the diffusion and use of ICTs are leading to both positive and negative environmental impacts.

4.2. Planned Obsolescence and the E-Waste Crisis

The idea that producers might want to decrease the durability of their goods in order to induce customers to replace their goods more frequently is consistent with the idea that 'reuse of products' or 'products with long lives' reduce demand for new products. Planned obsolescence strategies and other marketing tactics are practiced by the electronics industry, in an effort to increase consumer demand for newly developed products. Rapid technology change, low initial cost of manufactured goods, and even planned obsolescence results in a fast growing waste problem around the globe.

Table 3. Impacts of ICT advancement

Type of Impact	Notes on implication
Direct impacts	The direct impacts of the production and use of ICTs on the environment include resource use and pollution related to the production of infrastructure and devices, electricity consumption of hardware, electronic waste disposal etc
Indirect impacts	The indirect impacts related to the effect of ICTs on production processes, products and distribution systems include de-materialisation, substitution of information goods for material goods, and substitution of communication at a distance for travel etc; and
Structural/behavioural impacts	Structural/behavioural impacts, mainly through the stimulation of structural change and growth in the economy by ICTs, and through impacts on life styles and value systems.

For many products, environmentalists assume that reuse is environmentally beneficial because it replaces the manufacturing and purchase of new goods. However, manufacturers may oppose this type of reuse for the same reasons as this will impact negatively on production of new EEEs and sales.

Thomas (2003) observed that there is a rich economic literature on planned obsolesce, the incentives to producers to alter the durability of their products, and the circumstances that promote or inhibit second-hand markets. Factors that affect the reuse of products extend beyond durability *per se* and include manufacturers' practices with respect to product maintenance, access to spare parts, software upgrades and compatibility, and copyright protection (Thomas, 2003). The design and manufacture of a new model of say a printer while phasing out the production of spare parts for earlier brands would definitely influence negatively on the life and reusability of the products already with the consumer.

4.3. The Global Crisis of Electronic Waste Generation and Management

The term electronic waste (e-waste) or waste electrical and electronic equipment (WEEE) refers to unwanted electrical and electronic equipment (EEE) that are obsolete, at the end of their lives or that have been discarded by their original users. In general, "an electrically powered appliance that no longer satisfies the current owner for its original purpose' is viewed as e-waste. E-waste encompasses a broad and growing range of electronic devices ranging from large household appliances such as refrigerators, air conditioners, mobile phones, consumer electronics, computers etc. that have been discarded by their users.

Consequently, e-waste is the most rapidly growing waste problem in the world today. It is a crisis, not only of quantity, but also a crisis born from hazardous material content that pose both human and environmental health threat. E-waste has become a problem of crisis proportion because of two primary characteristics:

- The high rate of e-waste generation- E-waste is generated at alarming rates due to obsolescence.
- E-waste is hazardous- It contains over 1,000 different substances, many of which are toxic, and creates serious pollution upon disposal especially when inappropriate routes are adopted.

4.3.1. Crisis of Quantity

There is no doubt that the growth in electrical appliances and electronic equipment has been exponential, especially in the last 2-3 decades. The useful life of consumer electronic products is relatively short, and decreasing because of rapid changes in equipment features and capabilities (Kang and Schoenung, 2005). Due to extreme rates of obsolescence, e-waste produces much higher volumes of waste in comparison to other consumer goods.

The increasingly rapid evolution of technology combined with rapid product obsolescence has effectively rendered everything disposable. The average lifespan of most consumer electronic products have decreased dramatically.

For example, the lifespan of a computer has shrunk from 4-5 years to 2 years (Reference). Again, consumers now rarely take a faulty or broken electronics for repairs. This is because it is now often easier and cheaper to replace an electronic product than to repair it. Available estimates show that as much as 500 million computers reached their end-of-life worldwide between 1994 and 2003 (Widmer et al. 2005). This waste volume may contain as much as 2 870 000 tons of plastics, 718 000 tons of lead, 1363 tons of Cd and 287 tons of mercury (Widmer et al. 2005).

4.3.2. Crisis of Hazardous Material Contents

The valuable fraction in including iron, copper, aluminium, gold and other metals in e-waste is over 60%, while pollutants comprise 2.70% (Figure 3) (Widmer et al, 2005). The typical e-waste scrap consists of copper (20%), iron (8%), tin (4%), nickel (2%), lead (2%), zinc (1%), silver (0.2%), gold (0.1%), and palladium (Sodhi and Reimer, 2001). Computer CRTs present a disposal problem because of their growing magnitude in the waste stream and their role as a major source of Pb in MSW (Musson et al, 2000). CRTs in computer monitor and televisions contain about 8% lead by weight (Li et al, 2006; Smith et al, 1996); amounting to about 2- 4kg of lead (Powell, 2002). Consumer electronics account for 27% of Pb discards in MSW in 1986 in the US and was projected to make 30% of lead discards by 2007. Other substances of environmental health concern in e-waste include heavy metals such as cadmium, chromium, mercury, antimony and brominated flame-retardants (BFRs) such as polybrominated diphenyl ether (PBDE), and polybrominated biphenyls (PBB).

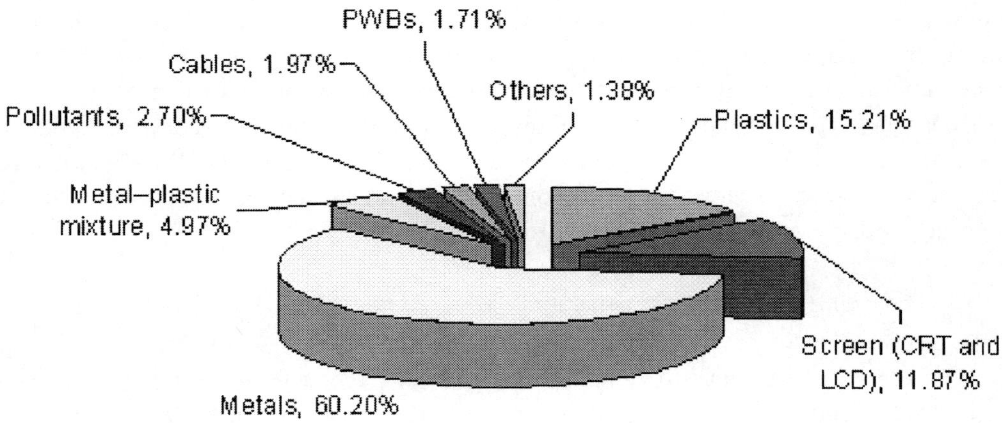

Figure 3. Material fractions in E-waste (as reported by a recycler in Switzerland).

5. Transboundary Movement of E-Waste – An Environmental Injustice

E-waste is routinely exported by developed countries to developing ones, often in violation of the Basel Convention (Greenpeace 2005b; 2008). Williams (2005) observed that few if any statistical categories are designed to distinguish new goods from used or waste ones.

5.1. Concern over E-Waste Exports

In an attempt to bridge the digital divide, there appears a need for the developed countries to assist the developing countries with ICT infrastructure and equipment among other requirements. In fact, Williams (2003) observed that domestic and international trade in used electronics presents a possible important avenue to stimulate the dissemination of information technology and bridging the gap between the "*haves*" and the '*have nots*".

As a result, large quantities of used electronics are exported to developing countries. The export of used electronics leaves much to be desired:

1. there is no distinction between items for reuse and items for recycling
2. there is no guarantee of the functionality of the exported devices,
3. some vital components of exported devices are often missing. Sometimes they may not, for example, have the correct plug fitted, or the right software installed. For example, some used keyboards for export may be in different language.
4. most appliances that do work on arrival only have a short second life, as they were already old, obsolete and / or damaged during the transit and,
5. it is not easy to distinguish between legal and illegal exports.

Consequently, increasing volumes of e-waste are being managed in developing countries. E-wastes are exported to developing countries despite laws at regional and global level prohibiting the export of hazardous e-waste. Some developed countries export considerable quantities of e-waste to China, India, Nigeria and Ghana and other developing countries, even though the Basel Convention restricts transboundary trade of it (Huang et al, 2009). The Basel Convention regulates the transboundary movement of e-waste as it is considered dangerous to humans and the environment under the List A of Annex VIII of the Convention (UNEP, 1989)

5.2. Origin of E-Waste Exports/Dumping

In the documentary *Toxic Tech: Pulling the Plug on Dirty Electronics*, Greenpeace revealed the origin of e-waste exports:

> "In the 1990s, governments in the European Union, Japan and some US states set up e-waste 'recycling' systems. However, many countries did not have the capacity to deal with the sheer quantity of e-waste they generated or with its hazardous nature. Therefore, they began exporting the problem to developing countries where laws to protect workers and the environment are inadequate or not enforced. It is also cheaper to 'recycle' waste in developing countries; the cost of glass-to-glass recycling of computer monitors in the United States is \$0.50 per pound compared to \$0.05 in China". (Greenpeace, 2005b)

The trade in e-waste is encouraged because it is cheaper to 'recycle' waste and costs little or nothing to dump or burn in developing countries. Yang et al (2007) observed that "in the importation business of e-waste, it is not easy to distinguish between legal and illegal importation. Illegal importation of WEEE from other countries exists, but there is no statistical data for the quantity of illegal imports of WEEE into China".

This appears to be the scenario in most developing countries Similarly, Townsend, (2006) reviewed the scene in the Americas and observed that

> "The transfer of personal computer equipment and other electronic devices from the US, Canada and similar countries to Latin American and the Caribbean countries offers a tremendous opportunity to reduce the digital divide between the North and the South. There is concern, however, that this practice could result in a trans-boundary shipment of wastes which may under some definitions be regulated as hazardous".

5.3. Statistics of Transboundary Movement of E-Waste

There have been documented evidences of the trans-boundary movement of e-waste around the world. For instance in Hong Kong, 40, 28 and 2 shipments of waste TV cathode ray tubes were intercepted from North America, Asia/Oceania, and Europe respectively in 2004 (Jefferies, 2006). An assessment of e-waste management in India indicated that about 7200 tonnes of e-waste are imported daily in Delhi alone (Dimitrakakis et al, 2006). Tasaki et al (2004) observed that an estimated 4.27 million of the 8.9 million (old) TV sets generated in Japan were exported. This accounted for about 48% waste TV sets generated in the country.

Williams (2005) reported that 160,000 metric tons of secondary and waste electronic equipment were exported from the UK In 2003. Of this, about 133,000 tons was IT/telecom equipment. In this category, 110,000 tons were declared exports and properly documented while 23,000 tons were undeclared or grey-market exports going to non-OECD countries including Far East, India, Africa and China. (Williams, 2005). The Greenpeace study (Greenpeace 2009b) reported that an inspection of 18 European seaports in 2005 found as much as 47 per cent of waste destined for export, including e-waste, was illegal.

Between 50 and 80% of e-waste collected for recycling in the USA each year are being exported, amounting to about 10.2 million PCs (Roman and Puckett, 2002; BAN/SVTC, 2002; BAN, 2005; Greenpeace, 2005b). Estimates show that South Korea exports an estimated 1.8 million units of used computers to China each year (Toxic Dispatch, 2004). Mainland China tried to prevent this trade by banning the import of e-waste in 2000. Unfortunately, Greenpeace observed that the laws are not working; e-waste is still arriving in Guiyu of Guangdong Province, the main centre of e-waste scrapping in China (Greenpeace 2005a).

6. Management Routes in Developing Countries: Environmental and Health Issues

E-wastes contain a complex array of hazardous substances that can threaten human health and the environment if not disposed properly. Due to their hazardous material contents, e-waste may cause environmental and health problems during the waste management phase if it is not properly pre-treated.

6.1. Management Practices: A Review of Guiyu Recycling Town

When introduced into the landfills and incinerator in the traditional waste stream, e-waste exposes humans and the environment to the hazardous material contained within the electronic products. Since e-waste has become one of the most rapidly growing components of the waste stream, its management is an important environmental and public health concern.

The major constraints in achieving sound management of WEEE in the developing countries are the absence of an organized reverse flow of EoL devices and the unavailability of formal recycling infrastructure (Osibanjo and Nnorom, 2007). In most developing countries, e-waste are managed through various inappropriate routes including disposal at open dumps, unsanitary landfills, and material recovery through back yard recycling (informal recycling). Most developed countries, in contrast, have adopted various policies initiatives and regulations for improved management strategies for WEEE management.

6.2. Human and Environmental Contamination from Inappropriate Management of E-Waste

6.2.1. Soil Contamination

A typical example of the hazards of crude recycling of e-waste is the Guiyu town (Chaozhou region) in southeastern Guangdong province of China. This town attracted so much international attention because of the impact of e-waste recycling activities. In this town, primitive recycling techniques used include open burning of plastics (to reduce waste volume), and copper wires (to salvage valuable metals e.g. copper); strong acid leaching of printed wiring boards, PWBs (to recover precious metals) from which the waste acid were discharged into nearby streams, and grilling of PWBs over honeycomb-coal fires to melt solder (to allow collection of electronic components, e.g. diodes and resistors) (Wong et al, 2007a; 2007b; Greenpeace, 2008; 2005a; 2005bLiu et al, 2006; Roman and Puckett, 2002).

These operations are usually carried out with no or very little personal protection equipment or pollution control measures. As a result, high levels of human exposure to toxins and environmental pollution have been reported (Wong et al, 2007; Liu et al 2006; Leung et al, 2004; Roman and Puckett, 2002; Greenpeace, 2005a; BAN, 2002). For example, polybrominated diphenyl ethers (PBDEs) levels in combustion residues from open burning in Guiyu were some of the highest ones found in any environmental medium (33 000 - 97 400 ng/g), and are more than 16 000 times higher than those found in soil samples in a distant reservoir that served as a control site (Leung et al, 2007; Huang et al, 2009). High levels of soil contamination with heavy metals have been reported at e-waste recycling sites (Greenpeace, 2005; Leung et al, 2004; Wong et al, 2007a). The soil samples from China contained up to 1.8% Cu and 0.16% Pb whereas the sites in India contained 1.1-1.4% Cu and 0.36-0.65% of Pb (Greenpeace, 2005a). These values far exceeded the critical limits for these metals in soils set by the EU and USEPA.

6.2.2. Water Contamination

River water sample from the Lianjiang and Nanyang rivers near the Guiyu "recycling village" in China indicated high levels of contamination with some heavy metals (Wong et al,

2007; Huang et al, 2009). The water samples revealed lead levels that were 2400 times higher than World Health Organization Drinking Water Guidelines. Sediment samples collected from the Lianjiang River also indicated high levels of lead levels about 212 times higher than what would be treated as hazardous waste (Huang et al, 2009).

The total PBDE concentrations ranged from 4434 to 16088 ng/g (dry weight) in Nanyang River bank sediment, from 55 to 445 ng/g in Nanyang River bottom sediment and 51.3 to 365 ng/g in Lianjiang River bottom sediment in Guiyu compared with those from 16.1 to 21.4 ng/g in wastewater discharged from a vehicle repairing workshop in Lo Uk Tsuen in Hong Kong (Luo et al, 2007).

The study also reported a mean concentration of total PBDEs in mixed muscles of tilapia (*Oreochromis spp*) from Lianjiang River were 115 ng/g wet weight and from wastewater in Hong Kong were 4.1ng/g wet weight (ww). Highest mean PBDE concentration was obtained in liver (2687ng/g ww), followed by abdomen muscle (1088ng/g ww) of bighead carp (*Aristichthys nobilis*) collected from Nanyang River. A significant correlation of concentration of each PBDE congener between sediment and muscle from Guiyu was observed. The present results of total PBDEs in sediment and fish were 10 and 1000 times higher than other studies. Open burning and dumping of e-waste are the major causes of PBDE contamination (Luo et al, 2007). Other studies indicate high-level contamination of the two rivers from the e-waste recycling activities (Wang et al, 2005; Wang et al, 2009).

6.2.3. Air Contamination

According to a report from State Key Laboratory of Organic Geochemistry in China, severe PCDD/Fs, PBDD/Fs, PBDEs and organochlorine pesticides (OCPs) pollution in air were investigated in Guiyu. The atmospheric PCDD/Fs abundances and toxic equivalent values were 64.9–2365 pg/m^3 and 0.909–48.9 pg of W–TEQ/m^3 respectively. These were the highest documented values of these compounds found in ambient air in the world. The total PCDD/Fs intake doses far exceeded the World Health Organization (WHO) 1998 tolerable daily intake limit of 1–4 pg of W–TEQk/g. To make matters worse, the severe dioxin pollution present in Guiyu substantially influenced the adjacent area of Chendian, where atmospheric PCDD/F and 2,3,7,8-PBDD/F levels were higher than those of common urban areas in the world. Local residents were also exposed to these toxic pollutants through inhalation, dermal exposure, and oral intake (Huang et al, 2009).

6.2.4. Human Exposure

Studies indicate that workers engaged in the material recovery from e-waste do so without personal protective devices (no masks or gloves) thereby exposing them to dangerous gases released, acid solutions/fumes, toxic smoke and contaminated ashes. For an intensive electronic waste recycling site in China with open burning it has been shown that the body burdens of people from these sites have significantly higher PCDD/PCF levels (human milk, placenta, hair) than those from reference sites. The estimated daily intake of PCDD/PCDFs for breast-fed infants has been found to be two times higher than for reference sites and to exceed the WHO tolerable daily intake 11 to 25 times (Chan et al. 2007).

Where soils and dusts are contaminated, children can be particularly exposed through hand-to-mouth transfer (Malcoe et al. 2002). This inadvertent oral ingestion of soil is considered an important exposure pathway for metals and other contaminants, especially in children through outdoor hand-to-mouth activities (pica).

It has been shown that children in Guiyu have significantly higher levels of blood lead than children living in other towns (Huo et al. 2007; Zheng et al, 2008).

7. Discussion (Finding a Global Solution)

7.1. Finding an International Solution

Finding a global solution to e-waste exports is dependent on the development of an international guideline for distinguishing between used goods for reuse and e-waste. Presently there are no internationally acceptable guidelines for the separate classification to differentiate used equipment meant for reuse and e-waste meant for recycling. Williams (2005) observed that few if any statistical categories are designed to distinguish new goods from used or waste ones. Hong Kong and Australia were identified as the first to develop guidelines for distinguishing between used goods and e-waste (Kojimo, 2005).

The international guideline should among others assure that equipment for reuse should:

1. be functional
2. have all vital components present
3. be of similar voltage/current capacity as is applicable in the receiving country
4. have a pre-determined life and possibly a warranty

Most stakeholders appreciate that used electronics or electronic waste should be viewed as a potential resource, rather than a waste burden. In this scenario, the following are critical:

- E-waste should be managed in ways that could reduce environmental impacts, and yield revenue via precious metals recovery;
- The sales, repair, use, and sound EoL management of e-waste would help improve local economy, and;
- The reuse of functional used EEE would help low-income areas bridge the *digital divide.*

The challenges in effective appreciation of the above are that the receiving countries have neither infrastructure nor policies for the EoL management of e-waste (Osibanjo and Nnorom, 2007). If appropriately applied the international trade in used electronics could be turned into a potential resource, rather than a waste burden. This can be achieved if only functional appliances are exported for reuse. In this scenario, the environmental impact of the exported ICT-wares will be reduced while bridging the digital divide (Townsend, 2006). In achieving this, Ladou and Lovegrove, (2008) suggested a "much tighter controls both on the transboundary movement of e-waste and on the manner in which it is recycled.

7.2. Managing E-Wastes Generated from Exports Used Electronic

The key components of extended producer responsibility (EPR) or product stewardship is the ownership of the EoL good by the key players- producer/retailer and/or the consumer. This could be in the form of paying an advance-recycling fee as enshrined in the EU WEEE Directive or the payment of recycling fee at the point of disposal of EoL products. Achieving global sound management of e-waste demands that manufacturers of electronic products take responsibility for their products throughout the products life cycle - from production to end-of-life. To achieve this Ladou and Lovegrove, (2008) suggested that manufacturers should focus on developing clean products with longer life-spans that are safe and easy to repair, upgrade, and recycle and will not expose workers and the environment to hazardous chemicals.

While bridging the divide with functional used electronics, provision should also be made preparatory to assuring effective management of these products on reaching their end-of-life. Some key issues to be resolved in finding solutions to EoL products, earlier exported as used products include:

1. Who monitors exports of used electronics to ensure compliance with regulations on the export of only functional products?
2. How should the 'exporting countries' assist the 'receiving countries' before, during and after EoL of these products?
3. How should the fees collected from consumers for appropriate e-waste recycling in the 'exporting' countries be allotted when these goods are exported rather than recycled at home? Who gets what?
4. What should be the role of the OEM in e-waste management when e-waste from EPR-compliant state and/or non- EPR-compliant states are exported to a developing country?

In simpler terms should the revenues collected for sound management of e-waste in the developed countries (EPR-compliant state) be transferred to the 'receiving countries' (non-EPR-compliant state) when the goods are exported to ensure that they are recycled using sound techniques at their end of life?

Should such funds be applied in assisting used EEE receiving countries with technology and infrastructure? If appropriately handled, the transboundary movement of functional used electronics could approach a win-win situation. As such, at the end-of-life of the handed down used EEE, the developed countries and manufacturers should also be at hand in finding an eco-efficient management option, which should include export back to the country of origin. In this scenario, the developed countries would assist in bridging the digital divide with only functional used ICT-wares at an affordable cost while at the same time supporting these countries with infrastructure and technology to prepare for challenges of managing these products in next few years. This would have both social-economic and environmental benefits. Public-private partnerships and intergovernmental cooperation are essential to finding solutions and addressing both the digital divide and the increasing global e-waste problems.

Conclusion

Access to ICT is critical in the development of any nation. Most developing countries are grappling with the challenges and constraints of the digital divide. Access to low cost new ICT-wares and functional used ICT-wares are important in bridging the digital divide if appropriately harnessed. A well-developed and formalized secondary market for used electronics is important in bridging the digital divide. There should however be distinction between 'equipment for reuse' - which should be tested and confirmed functional (using an internationally acceptable standard) - and 'electronic scrap' meant for recycling which should be treated as 'hazardous waste'. Such wastes should be exported as 'hazardous wastes' and only with the 'prior consent' of the receiving country which should also have established recycling facility. These activities would have both social and environmental benefits while decreasing digital divide, thus bridging the digital divide in a more sustainable way.

The major constraints in achieving sound management of e-waste in the developing countries are the absence of effective legislation and availability of formal recycling infrastructure. An internationally acceptable solution is required in solving the global e-waste problem. A key component of this is finding solution to the dumping of EoL electronics in developing countries. The dumping of e-waste in developing countries is environmentally unjust and undermines the ecosystem health. This practice is unsustainable and exposes developing countries to a cocktail of toxins thereby worsening their poverty. There is therefore a need for a global policy initiative and legislation in dealing with this emerging global crisis. The export of functional used ICT-wares would help improve local economies, help low-income areas bridge the digital divide while reducing the environmental impacts of the ICT-wares.

References

BAN. 2005. *The Digital Dump: Exporting Re-Use and Abuse to Africa*, the Basel Action Network. October 24, 2005. Jim Puckett (Editor). www.ban.org

BAN/SVTC. 2002. *Exporting Harm: the High Tech Trashing of Asia*. The Basel Action Network and Silicon Valley Toxics Coalition. February 25, 2002. Puckett, J and Smith, T (Eds.).

Berkhout F, Hertin J. De-materialising and re-materialising: digital technologies and the environment. *Futures* 2004;36:903–20.

Chan JK, Xing GH, Xu Y, Liang Y, Chen LX, Wu SC, Wong CK, Leung CK, Wong MH. 2007. Body loadings and health risk assessment of polychlorinated dibenzo-p-dioxins and dibenzofurans at an intensive electronic waste recycling site in China. *Environ. Sci. Technol.* 41:7668–74.

Dimitrakakis, E., Janz, A., Bilitewski, B. and Gidarakos, E., 2009. Small WEEE: determining recyclables and hazardous substances in plastics. *Journal of Hazardous Materials* 161:913-919.

Greenpeace 2005a. *Recycling of Electronic Waste in China and India: Workplace and Environmental Contamination*. Greenpeace Report. (Edited by Brigden, K., Labunska, I.,

Santillo, D., and Allsopp, M). Greenpeace International. August, 2005a. www.greenpeace.org.

Greenpeace 2005b. *Toxic tech: pulling the plug on dirty electronics.* Greenpeace International. Greenpeace Toxics Campaign May 2005. Published by Greenpeace International, Amsterdam, The Netherlands. Retrieved Aug. 13, 2007 from www.greenpeace.org.

Greenpeace. 2008. *Poisoning the poor: Electronic waste in Ghana.* Greenpeace Report, August 2008. www.greenpeace.org.

Hageluken, C. 2006a. Recycling of electronic scrap at Umicore's integrated metal smelter and refinery. *World of Metallurgy ERZMETALL* 59 No. 3 pp. 152-161. Paper originally presented on the occasion of the European Metallurgical Conference EMC 2005, September 18-21, 2005, Dresden vol. 1 pp. 307-323.

Hageluken, C. 2006b. Improving metal returns and eco-efficiency in electronic recycling – a holistic approach to interface optimization between pre-processing and integrated metal smelting and refining. In: *Proceedings of the 2006 IEEE International Symposium on electronics and the Environment*, 8-11 May, 2006, San Francisco. Pp. 218- 223.

Hilty, L.M., Arnfalk, P., Erdmann, L., Goodman, J., Lehmann, M. and Wager, P.A. 2006. The relevance of information and communication technologies for environmental sustainability- a prospective simulation study. *Environmental Modelling and Software* 21: 1618-1629.

Huang, K; Guo, J; Xu, Z 2009. Recycling of waste printed circuit boards: A review of current technologies and treatment status in China. *Journal of Hazardous Materials* 164: 399-408.

Huo, X., Peng, L., Xu, X., Zheng, L., Qiu, B., Qi, Z., Zhang, B., Han, D. and Piao, Z. 2007. Elevated blood levels of children in guiyu, an electronic waste recycling town in China. *Environmental Health Perspectives* 115: 1113–1117.

Jefferies, E. 2006. E-wasted: toys and gadgets become toxic junk, thanks to the circuit-bored. *Worldwatch* 19.4:21-25. July/Aug. 2006. Worldwatch Institute. www.worldwatch.org.

Jhin, P 2007. Collaboration Towards Information And Communications Technology For Development (ICT4D).December 12, 2007. www.gkpcms.com/GK3/documents/07.12.12-GK3-SS4-Paul%20Jhin.pdf.

Kang, H-Y. and Schoenung, J.M 2005. Electronic waste recycling: a review of U.S infrastructure and technology options. *Resources Conservation and Recycling* 45: 368-400.

Kojima, M. 2005. Towards appropriate resource recycling in Asia. *International trade of recyclable resources in Asia.* M. Kojima. Ed.

Ladou, J and Lovegrove, S 2008. Export of Electronics Equipment Waste. *Int. J. Occup. Environ. Health.* 2008;14:1–10.

Lall A and Garai A 2005. Capturing Grassroots Voices in the Information Society and Sustainable Development: The Policy Dialogue in India. In: A Developing Connection: Bridging the Policy Gap between the Information Society and Sustainable Development ISBN 1-895536-77-4. (Terri Willard, Maja Andjelkovic, Steve Vosloo, Wainaina Mungai, Margarita Salas, Anusha Lall, Atanu Garai, Diogo André de Assumpção, Amira Sobeih Eds), IISD, 2005. International Institute for Sustainable Development, www.iisd.org

Leung, A., Cai, Z.W. and Wong, M.H 2004. Environmental contamination from e-waste recycling at Guiyu, Southeast China. In: *Proceedings of the 3rd Workshop on Material Cycles and Waste Management in Asia*, Tokyo, 14-15 December, 2004, pp. 73-84.

Leung, A.W.; Luksemburg, W.J.; Wong, A.S.; Wong, M.H. 2007. Spatial distribution of polybrominated diphenyl ethers and polychlorinated dibenzo-p-dioxins and dibenzofurans in soil and combusted residue at Guiyu, an electronic waste recycling site in Southeast China, *Environ. Sci. Technol.* 41: 2730–2737.

Li, Y., Richardson, J.B., Walker, A.K. and Youn, P-C., 2006. TCLP heavy metal leaching of personal computer components. *Journal of Environmental Engineering* 132.4: 497-504.

Liu, X., Tanaka, M. and Matsui, Y. 2006. Electrical and electronic waste management in China: progress and the barrier to overcome. *Waste Management and Research* 24: 92-101.

Luo Q, Cai Z.W, Wong, M.H 2007. Polybrominated diphenyl ethers in fish and sediment from river polluted by electronic waste. *Science of the Total Environment* 383 (2007) 115–127.

Marine, S. and Blanchard, J.M. 2004. Bridging the digital divide: an opportunity for growth in the 21st century. *Alcatel Telecommunications Review- 3rd Quarter 2004. Strategy White Paper. Alcatel.* Pp. 1-8. www.alcatel.com

Monge, R and Chacón, F 2002. Bridging the Digital Divide in Costa Rica Access to and Use of Information and Communications Technologies (ICTs). Translation of the original version in Spanish, by: Xinia Rodríguez Castillo. Costa Rican Advisory Commission on High Technology (CAATEC). Digital Costa Rica Series 1. ISBN 9977-911-55-X.

Musson, S.E., Jang, Y-C., Townsend, T.G. and Chung, I-H. 2000. Characterization of lead leachability from cathode ray tubes using the toxicity characterization leaching procedure. *Environmental Science Technology* 34: 4376-4381.

Nnorom IC, Osibanjo O. 2008. "Overview of electronic waste (e-waste) management practices and legislations, and their poor applications in the developing countries", *Resources Conservation and Recycling* 52: 843-858.

Obijiofor, L 2009. Mapping theoretical and practical issues in the relationship between ICTs and Africa's socioeconomic development. *Telematics and Informatics* 26: 32–43.

Osibanjo O, Nnorom I.C. 2007. The challenge of electronic waste (e-waste) management in developing countries. *Waste Management and Research*, 25: 489-501.

Powel, J. 2002. *E-scrap future: recent key trends.* E-Scrap News. Resource Recycling Magazine. Regional Electronics Conference, Louisville, Kentucky, USA. Aug. 29, 2002.

Roman, L.S. and Puckett, J. 2002. E-scrap exportation: challenges and considerations. *Proceeding of the International Symposium on Electronics and the Environment 2002 IEEE.* May 6-9, 2002. San Francisco, CA, USA. 79-84.

Smith, D., Small, M., Dodds, R., Amagai, S. and Strong, T. 1996. Computer monitor recycling: a case study. *Engineering Science Education Journal* 5.4: 159-164.

Sodhi, M.S. and Reimer, B. 2001. Model for recycling electronics end-of-life products. *OR Spektrum* 23, 97–115.

Tasaki, T.; Tasasuga, T.; Osako, M.; Sakai, S.I. 2004. Substance flow analysis of brominated flame retardants and related compounds in waste TV sets in Japan, *Waste Manag.* 24: 571–580.

Thomas, V.M. 2003. Demand and dematerialization impacts of second-hand markets: reuse or more use? *Journal of Industrial Ecology* 7.2: 65-78.

Townsend, T 2006. Government challenges and opportunities in developing electronics recycling programs. Presented to: SUR Corporacion de Estudios Sociales y Educacion Santiago, Chile. Prepared by: Timothy Townsend, University of Florida. October 2006.

Toxic Dispatch. 2004. *Environmentalists Denounce Toxic Waste Dumping in Asia.* Toxic Dispatch No 23 Sept. 2004. A newsletter from Toxic Links. 1-2.

UNEP 1989. Basel convention on the control of transboundary movements of hazardous wastes and their disposal, United Nations Environment Programme/Secretariat of the Basel Convention; 1989. http://www.basel.int/text/documents.html.

Wang D, Cai, Z, Jiang G, Leung A, Wong MH., Wong, WK 2005. Determination of polybrominated diphenyl ethers in soil and sediment from an electronic waste recycling facility. Chemosphere 60: 810–816.

Wang, F; Leung, A.O.W; Wu, S.C.; Yang, M.S; Wong, M.H. (2009) Chemical and ecotoxicological analyses of sediments and elutriates of contaminated rivers due to e-waste recycling activities using a diverse battery of bioassays. *Environmental Pollution* 157: 2082–2090.

Widmer, R., Oswald-Krapf, H., Sinha-Khetriwal, A., Scnellmann, M. and Boni, H. 2005. Global perspectives on the e-waste. *Environmental Impact Assessment Review* 25: 436-458.

Williams, E. 2005. International activities on E-waste and guidelines for future work. *Proceedings of the third workshop on materials cycles and waste management in Asia*, National Institute of Environmental Sciences: Tsukuba Japan.

Williams, E.D. 2003. Extending PC lifespan through secondary markets. *2003 IEEE International Symposium on Electronics and the Environment.* May 19-22, 2003. pp 255-259.

Wilson, E.J. and Wong, K. 2003. Africa information revolution: a balance sheet. *Telecommunication Policy* 27: 155-177.

Wong CSC, Wu, S.C., Duzgoren-Aydin, NS., Aydin, A; Wong, MH. (2007) Trace metal contamination of sediments in an e-waste processing village in China. *Environmental Pollution* 145: 434-442.

Wong, C.S.C., Wu, S.C., Duzgoren-Aydin, N.S., Aydin, A. and Wong, M.H. (2007a). Trace metal contamination of sediments in an e-waste processing village in China. *Environmental Pollution* 145 (2): 434-442.

Wong, M.H., Wu, S.C., Deng, W.J., Yu, X.Z., Luo, Q., Leung, A.O.W., Wong, C.S.C., Luksemburg, W.J. and Wong, A.S. (2007b). Export of toxic chemicals: A review of the case of uncontrolled electronic-waste recycling. *Environmental Pollution* 149: 131-140.

Ya'u, Y.Z (2002). Confronting the digital divide: an interrogation of the African initiatives at bridging the gap. *Paper presented at the CODESRIA/TWN Conference on Africa and the Challenges of Development in the New Millennium.* Accra, Ghana. Apr. 2002.

Yang J, Lu B, Xu C. (2007) WEEE flow and mitigating measures in China. *Waste Management* 28:1589–97.

Zheng, L., Wu, K., Li, Y., Qi, Z., Han, D., Zhang, B., Gu, C., Chen, G., Liu, J., Chen, S., Xu, X.and Huo, X. 2008. Blood lead and cadmium levels and relevant factors among children from an e-waste recycling town in China. *Environmental Research* 108: 15-20.

In: E-Waste: Management, Types and Challenges
Editors: Yuan Chun Li and Banci Lian Wang
ISBN 978-1-61942-217-9

Chapter 4

WHAT DO E-WASTE QUANTIFICATION STUDIES OF THE PEOPLE'S REPUBLIC OF CHINA TELL US?

Chung Shan-shan*
Department of Biology, Hong Kong Baptist University,
Kowloon Tong, Hong Kong SAR, PRC

ABSTRACT

With increasing exposure of the harm from electronic and electrical waste (e-waste) recycling in the People's Republic of China, it is an imperative that its generation and recycling be sufficiently regulated. Accurately depicting the rate of generation and disposal of e-waste is an important first step in planning for adequate formal e-waste logistics and treatment capacities.

A critical review of the quantification studies on China's e-waste generation and disposal rates shows that there are large discrepancies in the predicted amounts of e-waste generated no matter the same or different methods of estimation were used. Moreover, even when agreements between studies are found, the agreed figures are not necessarily the correct figures. There appears to be an imminent need to increase the transparency of data collection and computation methods in these studies as the lack of such transparency is identified as the main reason for estimation uncertainty.

Evident from this chapter, irresponsible citation practices are found to spread to academic studies already. It is urged that leading organizations in the academic community should consider establishing a platform devoting to the reporting of false or dubious citation.

Keywords: Electrical and electronic waste; waste quantification; WEEE disposal; People's Republic of China, E-waste generation

* Tel: 852-34117741. Fax: 852-34117743. E-mail: sschung@hkbu.edu.hk.

1. Introduction

E-waste generated in developed countries is said to find its way into less developed nations such as China, India, the Philippines and Nigeria either through exploitation of legislative loopholes or simply illegally (Kellow, 1999; Li and Duan, 2008; Nnorom and Osibanjo, 2008). Some reports have suggested that about 50-80% of scrapped computers from the US are exported to (coastal areas of) mainland China (Puckett and Byster, 2002; Xinhua News Agency of China, 2006). Environmental and health problems emanating from recycling of electronic and electrical waste (e-waste) in some parts of the People's Republic of China (Leung et al., 2007; Wong et al., 2007) have triggered a number of studies covering technical, logistics and regulatory aspects of e-waste. In this context, one major aspect is the amount of e-waste generated in China, which is important information for planning for appropriate measures to mitigate the adverse impact of e-waste. A large number of studies were reviewed for preparing this chapter but only eight of them were deemed suitable for a detailed review. Interestingly, despite use of the same input databases and same quantification methods, results of one study can be very different from another. In this chapter, common pitfalls and the source of uncertainties that would lead to disagreements in findings of e-waste quantification studies are identified so that users of similar data from future studies are better informed of how reliable these data are. To fully understand why sources of uncertainties are present, a review of generic e-waste quantification methods is indispensable and is presented below.

2. Quantitative Estimation of E-Waste

Walk (2004) mentioned three methods for estimating waste quantities, namely, projections (also known as the time-series model), input-output models and factor models. Among the three, the input-output model is the most promising and, therefore, most frequently used to make quantitative estimations of e-waste. Two commonly used approaches that come under this input-output concept are "market supply methods" and "consumption and use approach". Both approaches assume a delay (equivalent to the lifespan of appliances) for electrical and electronic products (e-products) to become e-waste. Thus, assumptions or information of product lifespan are needed to complete the estimation. In addition to what Walk (2004) has stated, there is one seldom mentioned approach to quantify the disposal of e-waste (as opposed to generation). This is the direct waste analysis method. Needless to say, each approach or method has its own range of applications and limitations.

2.1. Market Supply Method and Material Flow Approach

A commonly used approach is to use domestic sales figures of e-products to project the generation rate by further assuming or obtaining the typical weight and lifespan of each type of e-waste. Since theoretically domestic sales figures exclude exports, no further correction by trade statistics is needed. This is known as the market supply method. This approach is used by Feszty et al. (2003), Cobbing (2008), Matthews et al. (1997), Müller et al. (2009), Liu et al. (2006a, 2006b) and the Asia-Pacific Regional Centre for Hazardous Waste Management Training and Technology Transfer (ARCTTT) (2005). In some literature, it is also called the material-flow approach which uses annual domestic production plus imports minus exports to depict local consumption (Hsu and Kuo, 2005). The material-flow approach has been used by the US Environmental Protection Agency (EPA) to assess general waste composition. Two earlier German studies (Lohse et al., 1998) and many more recent studies have used the material flow approach to quantify e-waste and make projections (USEPA, 2002; Steubing, et al., 2010; Yang et al., 2008, Aizawa et al., 2008; Yoshida et al., 2009).

Both the market supply method and the materials flow approach are capable of accurately estimating the amount of e-waste generated because domestic sales or production figures (save unreported production or imports or exports) directly reflect the quantum of e-products that eventually get converted into waste. However, obtaining accurate estimates of e-waste disposal rate also depends on the availability of e-waste exports and domestic recycling and reuse rates. The last two pieces of information are rarely available in economies not having official e-waste take-back schemes. In some studies, even export and import figures were ignored (see for instance, Matthews et al., 1997). Moreover, many free-trade economies lack complete production information and accurate trade statistics (e.g., Hong Kong Special Administrative Region [SAR]). This method alone also cannot be used to estimate e-waste disposed for a certain year in an accurate manner.

In order to forecast the generation rate of e-waste for a specific year by either the market supply or the materials flow method, some distribution model has to be assumed. By applying the S-shaped curve model, using historical sales data on computers and making assumptions of computer lifespan, two studies (Yang and Williams, 2009; Yu et al., 2010) have forecasted computer waste generation for specific years in the future. Others prefer using a specific type of S-shaped curve, the Gompertz curve (ARCTTT, 2005), or other distributions, such as the Gaussian or normal obsolescence rate distributions, to forecast computer and other e-waste generation in specific regions, for specific years (Müller et al., 2009).

2.2. Direct Analysis on the E-Waste Stream

The second approach is to obtain direct information on the amount of e-waste disposed or processed. This is called direct waste analysis and can be done in two ways. First, records of e-waste receipts in all local waste facilities are to be obtained to estimate the total quantity of e-waste disposed from a defined community. An example of this can be found in Florence and Price (2005). However, one drawback with this approach is that e-waste receipt records from all waste facilities are not necessarily available in all jurisdictions. As an alternative, historical records from e-waste processors can also be used to compile macro-scale data in accordance with the principles behind this approach. The Californian Integrated Waste Management Board used this method to estimate quantities of cathode ray tube (CRT) televisions, computer monitors and central processing units diverted from the waste stream for refurbishment, repair or dismantling (2001).

Second, direct e-waste disposal data can also be obtained by field determination in waste facilities. Waste samples are taken at major end-disposal facilities or other facilities whenever appropriate (e.g., refuse transfer stations) and quantities of relevant e-waste are categorized and (preferably) weighed. This is the method the Hong Kong SAR government has used to estimate e-waste disposal figures of the city (Environment Bureau, 2010). Among the three approaches reviewed in this chapter, this is the only one capable of providing an estimate of e-waste disposed without additional survey (see Table 1). While direct and exact, this approach is costly to conduct and has high health and safety risk and a large coefficient of variation owing to difficulties in selecting representative samples because e-waste is not a frequently appearing waste material. E-waste generation (as opposed to disposal) figures and projections of future disposal rates cannot be extrapolated directly from this method but can be obtained in combination with other waste projection methods such as factor based models or time-series models (Chung, 2010).

2.3. Consumption and Use Approach

The third approach is the "consumption and use" approach which is also known as the "Estimate Method" (Wilkinson et al., 2001). Similar but more mathematically-oriented variants of this approach are the "use-phrase analysis" and the time step method (Lohse et al., 1998; Walk, 2004). On a macro scale, e-waste generation rate is worked out by dividing total consumption (or existing stock) of e-products of an economy by the average lifespan of the e-product in question. The European Environment Agency (2003) used this method to project the quanta of waste from refrigerators, televisions and PCs for some countries in Europe. In addition, Brazil has also used this method to estimate its computer waste (Schluep, 2009). Macro level data can also be extrapolated from micro level e-product consumption data when e-waste generation rate from a representative family or an individual is readily available. If not, then separate studies, such as questionnaire survey, have to be undertaken to obtain such information. The same as the market supply method, assumptions about or modeling of the average weight and lifespan of each type of e-waste of interest are still needed.

Walk (2009) forecast quantities of discarded CRT appliances in a German state by deploying available data on CRT ownership per household. The household-based figures and trends were then extrapolated to derive e-waste generation rate for the entire economy (Lohse et al., 1998). Bureau B&G also used this approach to obtain e-waste figures for the Netherlands (Lohse et al., 1998). While this method is suitable for making estimates of both e-waste generated and to obtain data of household inventory of discarded e-products or e-waste (which will eventually enter the waste stream), it is relatively costly to obtain representative household e-product ownership figures which are generally not readily available (Chung et al., 2011).

In addition, if such household-based data are to be obtained from questionnaire survey, responses from the respondents are more likely to be ballpark values only (Chung et al., 2011). Thus, it is often the last resort when no other reliable methods are feasible. If the more mathematically oriented variants are used, problems in obtaining data are even greater.

While most studies have used one single method to estimate e-waste in order to maintain consistency, some have deployed multiple approaches owing to data deficiency. Wikinson et al. (2001) is a typical example; variants of both market supply and consumption are used to derive a range of waste quantities for six types of e-products. Formerly, a United Nations Environment Programme study concluded in 2009 that waste from personal computer (PC) is the most studied type of e-waste (Schluep et al., 2009).

Table 1. Pros and cons of e-waste quantification methods

	Generation	Disposal	Merits	Major Shortcomings
Market supply/ materials flow	✓	×	Has similar level of accuracy as the sales or production data themselves Covers the whole economy A long period of estimation can be covered High reproducibility	Produce only generic figures but not year specific figures unless there is further assumptions about the distribution of future e-waste Needs representative lifespan and average weight figures or distribution Other essential data are often not available or hard to obtain
Waste facility record compilation	×	✓	High accuracy in data measurement Full coverage of the waste stream Low cost Able to estimate disposal rates	Data are often not available Low reproducibility No waste projection possible by itself
Direct waste analysis	×	✓	Direct and precise measurement can be obtained High accuracy in data measurement Able to estimate disposal rates	Involves substantial health & safety risk Costly Difficult to ensure representative samples Low consistency No waste projection possible by itself
Consumption and use	✓	✓×	Detail end-user's information and their preference can be obtained Covers the whole economy A long period of estimation can be covered Both generation and disposal data can be obtained if consumer disposal survey is conducted as well.	Costly to obtain representative samples of adequate size Only ballpark values can be obtained Low reproducibility

Table 2. Estimations of e-waste generation rate in China

	Arisings (in Mt or Mu)	Scope	G/D	Year covered	Distribution model used	Methodology & source
1	5.88 Mu 20.41 Mu 5.78 Mu 10.48 Mu 6.76 Mu	PC (including notebook computers), TV, Refrigerator, Washing machine, Air conditioner	G	2006	n.s.	Consumption and use approach; Duan et al. (2007)
2	48.92 Mu 37.93 Mu 9.67 Mu 11.59 Mu 12.35 Mu	PC (including notebook computers), TV, Refrigerator, Washing machine, Air conditioner	G	2010		
3	10.73Mu 40.88 Mu 11.12 Mu 8.00 Mu 2.81Mu 73.54 Mu	PC, TV, Refrigerator, Washing machines Air conditioners Total	G	2006	Age distribution for PC only	Market supply method; Yang et al. (2008)
4	19.57 Mu 55.73 Mu 11.87 Mu 12.61Mu 5.50 Mu	PC, TV, Refrigerator, Washing machines Air conditioners	G	2010		
5	4.48Mu 33.5 Mu 9.76Mu 7.56Mu 0.65Mu	PC, TV, Refrigerator, Washing machines Air conditioners	G	n.s.	n.s.	Market supply method; Streicher-Porte and Yang (2007).

	Arisings (in Mt or Mu)	Scope	G/D	Year covered	Distribution model used	Methodology & source
6 [a]	10 Mu 32 Mu 14 Mu 16 Mu 2 Mu	PC, TV, Refrigerator, Washing machine, Air conditioner.	G	2005	n.s.	Market supply method; Liu et al. (2006)
7 [a]	10.5 Mu 41 Mu 10.5 Mu 7 Mu 2.5 Mu	PC, TV, Refrigerator, Washing machine, Air conditioner	G	2006		
8 [a]	19.5 Mu 55 Mu 11 Mu 12 Mu 5 Mu or 2.2 Mt	PC, TV, Refrigerator, Washing machine, Air conditioner. Total	G	2010		
9	5 Mu, 4 Mu, 6 Mu	TV, Refrigerator, Washing machine	n.s.	n.s.	n.s.	Methodology not reported, CBCSD (2007)
10	0.3 Mt 1.35 Mt 0.495 Mt	PC TV, Refrigerator	G	2007	n.s.	Direct citation and derived from the consumption and use approach; Schluep et al. (2009)
11	6.6 Mu	PC	G	2005	Gaussian	Market supply method; Müller, et al. (2009)
12 [a]	n.a. 40 Mu 14 Mu 15.5 Mu 30 Mu	PC, TV, Refrigerator, Washing machine, Air conditioner.	G	2005	Gompertz	Market supply method; Asia-Pacific Regional Centre for Hazardous WasteManagement Training and Technology Transfer (2005)
13*	300 Mu 52 Mu 15.5 Mu 16.5 Mu 42 Mu	PC, TV, Refrigerator, Washing machine, Air conditioner.	G	2010		

Table 2. (Continued)

	Arisings (in Mt or Mu)	Scope	G/D	Year covered	Distribution model used	Methodology & source
14	4.1 Mu 27.1Mu 6.0 Mu 12.5 Mu 9.7 Mu	PC, TV, Refrigerator, Washing machines Air conditioners	G	2005	n.s.	Market supply method; Wang et al. (2010)
15	6.7 Mu 35.0 Mu 7.7 Mu 12.1 Mu 11.6 Mu	PC, TV, Refrigerator, Washing machines Air conditioners	G	2006		
16	59.7 Mu 51.6 Mu 10.6 Mu 16.0 Mu 31.4 Mu	PC, TV, Refrigerator, Washing machines Air conditioners	G	2010		

Keys: Mt – million tonnes; Mu – million units; G – waste generated; D – waste disposed; n.s. – not specified; n.a. – not available.

a- E-waste figures are read from the graphs of the reference.

Reviews of a number of recently published studies make it evident that this trend has persisted. Table 1 is a summary of the pros and cons of the four variants from the three methods.

3. E-Waste Generation in China

Since ownership and sales figures of major e-products are routinely reported in China's national statistics, many researchers have used them to estimate e-waste generation in China. Table 2 summarizes findings from select studies using various methods for estimating e-waste quantitatively.

3.1. Selection Criteria for this Review

While there are numerous studies or reports mentioning the amount of e-waste in mainland China, many are secondary data only and the methodological steps, as well as the assumptions made for computing those figures, have not been explained in detail. This is particularly the case with figures cited in newspapers and websites. Without knowing the methodology used, it is not possible to compare the performance of different methodological approaches in e-waste estimation objectively. As a result, these dubious e-waste quantity estimates are not mentioned in this chapter, with the exception of one which is discussed for illustration purposes.

China Business Council for Sustainable Development (CBCSD) (2007), quoting the official Xin Hua News Agency, reported that every year about 15 million units of televisions, refrigerators and washing machines were discarded in China (CBCSD, 2007). In CBCSD's report (2007) neither the methodology, nor the year covered, nor the nature of the waste (generated or disposed) is clearly stated. Neglecting the data computation methodology and just reporting figures from other sources is a common practice; this is an irresponsible act (citation) and should be discouraged. As a result, this chapter will only cover studies in which sufficient detail about the databases used and methodological approach are reported. In addition, since the focus of this chapter is on e-waste quantities in mainland China, e-waste projection studies on a single Chinese city would not be compared. Thus, the study by Liu et al. (2006b) who project e-waste up to 2020 for Beijing and another by Chung et al. (2011) who estimate the amount of e-waste for Hong Kong are excluded from this chapter.

A total of eight studies are compared in this chapter. Apparent from Table 2 is that the market supply method is the most frequently used approach in estimating e-waste generated in mainland China. Despite the convergence in methodology adopted, discrepancies between findings of different studies are huge. In the following sections, comparisons of e-waste estimation findings are presented and attempts are made to explain the discrepancies.

3.1. Methods of Comparison

Since time periods and generic e-waste types covered by the eight studies are not standardized, in order to make legitimate comparisons, only findings from studies covering the same period(s) and the same genre of e-waste are discussed. Two fundamental approaches are used to make comparisons. Pairwise comparison is used when the two studies reviewed cover the same e-waste types or the same period. In cases where more than two studies are relevant, one study is used as the focal point of discussion.

Results from some studies cannot be compared with each other. Müller et al. (2009) and ARCTTT (2005) are the examples. While both have used the market supply method and have attempted to estimate e-waste generated in 2005, the former chose to use the Gaussian distribution model while the latter opted for Gompertz distribution; if their results can be rendered comparable, the conclusion would have a high chance of throwing light on which distribution model is more appropriate for e-waste estimation. This, however, cannot be done because the scope of coverage of the two studies has no commonalities: ARCTTT (2005) does not provide an estimate of PC waste for 2005 and Müller et al. (2009) only cover PC waste. Comparison between Schluep et al. (2009) and other studies also cannot be made without making arbitrary assumptions on unit weights of different e-waste types because Schluep et al. (2009) have provided estimates denoted in weight only for the year 2007 whereas other studies have focused on the number of units of e-waste generated.

3.2. Comparison of Findings Obtained by Market Supply Methods

Studies covered in the following sub-sections include Yang et al. (2008), Liu et al. (2006a) Streicher-Prote and Yang (2007), Müller et al. (2009), Wang et al. (2010) and ARCTTT (2005).

3.2.1. PC Waste Estimates Compared

Only three studies (Müller et al., 2009; Liu et al., 2006a; Wang et al., 2010) covered estimation of e-waste generated in 2005. There are two main differences in these studies. First, in Müller et al. (2009) and Liu et al. (2006a) PC sales data were used to make estimations while Wang et al. (2010) made use of PC production data to make the projection of PC waste generated.

Second, Müller et al. (2009) used the Gaussian distribution to make time-series estimates while Wang et al (2010) and Liu et al. (2006a) did not state the kind of distribution used in their studies. The result is that the estimated amount of PC waste in China in 2005 by Wang et al. (2010) was the smallest (see Table 3).

Table 3. The quantities and assumed lifespan of PC waste in 2005 for three studies

	PC waste in 2005 (million units)	Lifespan assumptions for PC (yr)
Wang et al. (2010)	4.1	6
Müller et al. (2009)	6.6	4.91 (σ=1.2)
Liu et al. (2006a)	10	3.5

Since the general principle is that an assumed shorter lifespan means more waste generated in the future, the discrepancies may be explained by the differences in PC lifespan assumptions. In addition, Müller et al. (2009) state clearly that they consider also waste from stored unused PCs while Liu et al (2006a) and Wang et al. (2010) were silent on whether waste from idle machines was included in their estimates.

Regarding the transparency of methodology, while Müller et al. (2009) described in detail the projection methods used, unlike Wang et al. (2010) the authors did not state the actual data series used as inputs to their models. Since in Liu et al. (2006a), e-waste projection data were stated only in graphs, some reading errors are inevitable and this also makes exact comparisons difficult. Nevertheless, it is still probable that the difference in assumed lifespan is responsible for the wide range of estimation results.

3.2.2. Comparison of Estimates for 2006

Three studies are relevant, namely, Liu et al. (2006a), Wang et al. (2010) and Yang et al. (2008). While the same market supply method were used by all researchers, Wang et al. (2010) as explained previously used production data of e-products as the basis of projection while Yang et al., (2008) and Liu et al., (2006a) used sales data which were defined as domestic production plus importation as input data of their model.

Among the three studies, Liu et al. (2006a) and Yang et al., (2008) have the highest degree of convergence with respect to all categories of e-waste compared and other than the assumptions of PC lifespan, their assumptions on the lifespans of TV sets, refrigerators, washing machines and air conditioners are the same. However, Liu et al. (2006a) have not differentiated clearly between the two sets of statistics, namely, data of possession of appliances by households and annual sales data for household appliances; it is not clear as to which set was used for computation and which was used as adjustment or other purposes. Yang et al. (2008), on the other hand, stated clearly that their model was based on product sales data of appliances. However, both studies are silent on which distribution model was used for projecting e-waste into the future. Thus, even though both studies reached similar conclusions, in terms of transparency and methodology details provided, Yang et al. (2008) appears to be a more convincing study. Since Yang et al. (2008) deemed a better research study than Liu et al. (2006a), I will compare Yang et al (2008) with Wang et al (2010) in this section.

While Yang et al. (2008) projected greater amount of PC, TV and refrigerator waste but smaller amount of washing machines and air conditioner waste than Wang et al., (2006a), the biggest difference lied in the waste quantities of air conditioners (second row of Table 4).

Table 4. The quantities air conditioner waste in 2006 and explanatory factors

	Yang et al. (2008)	Wang et al. (2010)
Air conditioner waste in 2006 (million units)	2.81	11.6
Domestic sales/industrial output of air conditioners in 2003 (million units)	11.8	48.2
Lifespan assumed (years)	10	8

This difference however cannot be fully explained by just a two-year difference in lifespans assumed for air conditioners (see Table 4). A more convincing explanation will be the set of input data used for the projections. The 2003 figure on which Wang et al. (2010) based are much higher than that used by Yang et al. (2008). It is therefore natural that Wang et al. (2010) come up with a very different projected waste value from Yang et al., (2008).

3.2.3. Comparisons of Waste Electrical Appliances for 2005 and 2010

Estimates of waste televisions, refrigerators, washing machines and air-conditioners for 2005 and projections for 2010 were provided by Liu et al. (2006a), Wang et al. (2010) and ARCTTT (2005). Since it has been demonstrated that findings of Liu et al. (2006a) and Yang et al. (2008) have high degree of agreement with each other and Yang et al (2008) was discussed in the last section, to save space and repetition, comparison with only Liu et al. (2006) is discussed in this section.

Between the 2005 e-waste projection of Liu et al. (2006a) and Wang et al. (2010), Liu et al. (2006a) came up with greater numbers of e-waste in general with the exception of air conditioners which was a low 2 million units. Since it has been shown in s.3.2.2 that a major reason for the difference in estimations was due to the deployment of sales statistics as the basis of estimation by Liu et al. (2006a), this discrepancy between Liu et al. (2006a) and Wang et al. (2010) will not be explained further in this section. Next comes the comparison between ARCTTT (2005) and the other two studies.

Estimates of waste generated from refrigerators and washing machines provided by Liu et al. (2006a) and ARCTTT (2005) are similar or largely similar for 2005 and 2010. There is a large discrepancy (about 8 million units) between the two in the case of televisions for 2005 with Liu et al.'s estimate (2006a) smaller than that computed by ARCTTT (2005). However, the two studies still agree regarding the quantum of television waste in 2010.

Again, the greatest difference is found between their estimates of air-conditioners waste. ARCTTT (2005) estimated that 30 million units of air-conditioners waste was generated in 2005, which was 15 times the estimate of Liu et al. (2006a) and three times that of Wang et al. (2010). ARCTTT's projection (2005) for air-conditioner waste for 2010 was also high but the gap is smaller when compared to the 2010 air-conditioner waste projection made by Wang et al. (2010). It is not clear as to why such big differences occurred in estimates of only air-conditioners. Even more puzzling is the fact that the assumed lifespan used for air-conditioner in ARCTTT (2005) was a year longer (11 years compared to 10) than Liu et al. (2006a) and three years longer than Wang et al. (2010). As mentioned previously, a shorter lifespan result in higher e-waste estimation. Therefore, the lifespan factor cannot explain the differences in the quanta of air-conditioner waste in 2010.

The use of Gompertz curve in modeling e-waste growth by ARCTTT (2005) and the use of different statistical bases for model input are likely to be the most convincing explanations for the major discrepancy in air conditioner waste predictions.

It is evident from Figure 3 in Liu et al. (2006a) that from 2003 – 2010, the rate of increase in air conditioner waste was very slow. On the other hand, ARCTTT (2005) projected figures from 1990, using the Gompertz distribution, resulting in 2005 being a period of rapid growth. In addition, ARCTTT (2005) used national trade and production statistics as inputs while (as previously mentioned) Liu et al. (2006a) may have made unreported manipulation of the data for model input. This may also be responsible for the surprisingly large discrepancy. The use of the Gompertz growth model can also be used to explain why ARCTTT (2005) made the highest waste projection for PC in 2010.

Thus, the growth distribution assumed, time horizon of projections and the existence of hidden computation methods are all plausible reasons for the large difference in estimates of air-conditioner and other waste generated in China. Since Wang et al. (2010) do not state which growth distribution model is used or other computation details, it is not possible to offer more explanations for the differences in study results.

3.2.4. Comparison between Year Specific and Non Year Specific Estimates

Streicher-Porte and Yang (2007) is the focus of discussion in this section. Among all studies, estimates of waste of PCs and air-conditioners made are the lowest or among the lowest in this study but its estimates of quantities of waste televisions, refrigerators and washing machines are similar to those computed by Liu et al. (2006a) and Yang et al. (2008), but lower than those worked out by ARCTTT (2005). On the whole, Streicher-Porte and Yang (2007) tend to give more conservative (smaller in quantities) estimates despite having assumed the same e-product lifespans as Liu et al. (2006a) and Yang et al. (2008). Similar to Wang et al., (2010), the computation method is not described in detail in Streicher-Porte and Yang (2007). Thus, it is not possible to explain such differences. In general, estimates of non-specific years have assumed that current e-products are rendered obsolete at an average rate and, therefore, a relatively short period of validity (less than 5 years) for the projection is assumed.

3.2.5. Comparison of Findings Obtained by Different Estimation Methods

Duan et al. (2007) is the only study that has used the consumption and use approach. Therefore, it is the focus of discussion in this section. However, no comparison can be made between Duan et al. (2007) and Schluep et al. (2009) because the years covered and the unit of waste are different, despite both having adopted the same quantitative estimation approach.

Among studies reviewed, findings of Duan et al. (2007) are more comparable with Yang et al. (2008) and Wang et al., (2010) as they covered the same years and the same generic types of e-waste. However, Duan et al. (2007) presented separate estimates for sub-categories of e-waste, including plasma display panel (PDP), liquid crystal display (LCD) televisions and notebook computers, in addition to traditional CRT televisions and desktop computers. In comparison, all other studies mentioned in this chapter have been ambivalent about whether notebook computers, PDP and LCD televisions were covered in their study. In order to render them comparable, it has to be assumed that estimations of television waste in Liu et al. (2006a) and Wang et al. (2010) also included PDP and LCD televisions waste and their estimations of computer waste also included notebook computers.

Under this assumption, it is concluded that Duan et al. (2007) tended to produce lower figures of e-waste than Yang et al. (2008) with the exception of air-conditioners for both 2006 and 2010 and computers for 2010 (Items 1-2 and 7-8 in Table 2). Yet, the quantities and growth rates of e-waste worked out by Duan et al. (2007) are largely similar (but generally smaller) to those in Wang et al. (2010) (Items 1-2 and 15-16 in Table 2). It is also apparent from the estimated figures that a low growth rate is asserted by Yang et al. (2008) for computer products waste. Lifespans assumed by Duan et al. (2007) for desktop and notebook computers are 6 and 5 years, respectively, while in Yang et al. (2008), an age distribution model is used to simulate the shortening of lifespan of computers in China from 1993 – 2003

(5.2 years to 3.5 years) and Wang et al. (2010) assumed 6 years of lifespan for all kinds of computers. As a result, lifespans assumed by Duan et al. and Wang et al. are longer than those used by Yang et al. and, therefore, the difference in lifespans cannot be used to explain the much higher PC waste projection made in Duan et al. (2007) and Wang et al. (2010) for 2010.

As a conclusion, no explanation of why and how Duan et al. (2007) and Wang et al. (2010) assumed such high rates of growth for computers can be given because the distribution model used for making projections have not been stated. However, excepting computer waste, it appears that the consumption and use approach tends to give more conservative (i.e. smaller quantities) estimates than the market supply method. This may be due to the fact that aggregated national sales data are likely to include a large number of appliances that would not be used locally.

3.3. Finding the Best Estimates

In sum, it is apparent that the most liberal estimates are several times the most conservative ones with respect to computers and air-conditioners waste. If the ARCTTT (2005) estimate of PC waste in 2010 is accurate (300 million units), each Chinese national would be producing 0.22 units of PC waste every year. This translates into a per capita PC waste generation rate in China being almost twice that of the more developed Hong Kong SAR, which has a PC waste generation rate of only 0.12 unit (figure derived by the author from Chung et al., 2011).

Similarly, ARCTTT (2005) has made very liberal estimates of air-conditioners waste generation. ARCTTT estimated that by 2010, there would be about 42 million units of air conditioners waste generated, almost eight times estimates of Yang et al. (2008). However, the question is whether the ARCTTT (2005) estimates (42 million units of air conditioners waste by 2010) are necessarily wrong.

In an attempt to answer this question, the author made an abridged calculation using the consumption and use approach, deploying the latest published national statistics of household appliances ownership. In 2008, each household in mainland China owned 1.0028 units of air conditioners (China Statistics Bureau, 2009). In that year, China had 420.3 million households (China Statistics Bureau, 2009). Thus, total number of air-conditioners owned in 2008 should be 421.4 million units and since air-conditioners are assumed to have a 10 years lifespan (Yang et al., 2008; Liu et al., 2006), it is expected that on average there would be about 42.1 million units of air-conditioners waste generated each year. Thus, while ARCTTT estimates are much larger than all other studies reviewed, it may in fact be a more realistic figure indicating that it may be appropriate to use an "S" distribution to model e-waste of China in the next decade or so.

Thus, the seemingly better and more vigorous studies may not provide the most accurate estimates of e-waste generation and the Gompertz growth model appears to be reasonably accurate in depicting the e-waste trend for China from 1990 - 2015. Indeed, it is hard to state definitively which study is more accurate with respect to all types of e-waste quanta. The safest conclusion appears to be that no studies can be deemed reliable for all estimations it provides.

4. Precautions in Data Interpretation

A number of problems were observed in the e-waste literature reviewed. These problems are noteworthy not just for the purpose of finding an accurate estimate of the amount of e-waste generated in China but also because similar problems may be found in reports on e-waste for other places as well.

A fundamental problem encountered in this review was that many of the figures of e-waste are secondary data, i.e. not derived or worked out by the authors but cited from other sources. While citation is a very common practice in newspapers and even official publications, it should be done only when the authenticity of the data can be assured or when the data are critically reviewed. Citing important figures without verification or review is an undesirable practice and errors are difficult to discover. Thus, this practice should not appear in academic journals or reputable reports. Unfortunately, self-discipline of some researchers has not been found at a sufficiently high level to stop this from happening. An example is the citation of China's computer waste generation figure in Nnorom and Osibanjo (2008), which stated that estimation of 4 million units of computer waste per annum for mainland China was based on a report published by Greenpeace (2005). In the Greenpeace (2005) report, the same computer waste generation figure was cited from UNEP (2005), which contained no personal computer waste estimation for China at all. Such practice (relying predominantly on secondary sources of data) has attenuated the reliability and creditability of academic studies. Owing to the dubious nature of their figures, Nnorom and Osibanjo (2008) and Greenpeace (2005) studies have not been included in this discussion.

Second, appropriate primary data on which e-waste estimates are based can be difficult to secure. These primary data include, but are not limited to, statistics on imports, exports, domestic consumption and product sales. While national sales statistics of e-products are available for mainland China, Feszty et al. (2003) remarked that regional e-product sales data for Scotland were not available and had to be derived from those of U.K. Similarly, municipal level e-product sales data are not available for Hong Kong, one of the special administrative regions of China. Even for statistics that are available, data integrity can be doubtful. Wilkinson et al. (2001) considered e-product sales data from Dublin's Central Statistics Office unreliable and did not use them directly for e-waste calculation for Ireland. This is particularly the case with external trade statistics which are usually compiled from information declared on trade declaration forms, and in some cases, only when shipment value exceeds a certain level would the trade data be published in official statistics (Crowe et al., 2003). Such data are invariably self-reporting in nature and seldom verified. A case in point is the doubtful reliability of export data of television sets in Hong Kong (Chung et al., 2011).

Third, it is apparent from the above comparisons of studies adopting the market supply method that even if the same raw data are used and lifespans of e-products are standardized, this does not guarantee agreement in results. So, does it mean that the market supply method does not generate reproducible results? As mentioned above, the lack of agreement among studies using the same quantitative estimation approach may stem from incomplete or unclear descriptions of methodological details rather than the methodology itself. As a result, researchers are urged to always adopt the good practice of reporting their methodologies in full so that policy makers and facility planners have all the information they need to decide which studies are more reliable or more useful to them. Fourth, readers must be reminded that waste generation figures do not necessarily reflect quantities of waste disposed. However, waste generation figures are important figures for planning of collection, recycling and treatment infrastructure. E-waste disposal figures, on the other hand, indicate, among others, the scale of resource wastage and represent the upper boundary of the recovery potential. Nevertheless, no studies reviewed in this chapter are able to provide an estimate of quantities of e-waste disposed in China. Thus, there is a serious lack of information on the recovery potential of e-waste in China.

Conclusion

This review shows that PC is the most studied type of e-waste. A former review (Schluep, 2009) and this review also confirm that washing machines and refrigerators waste is the least covered waste type in available literature. It is also apparent from the present review that irresponsible citation practices have spread even to academic studies, irrespective of whether the published articles or studies have undergone the peer-review process. To deter irresponsible citation behavior, it is suggested that leading organizations in the academic community establish a publicly accessible platform for reporting of false or dubious citations to reduce the adverse effects on credentials of academic works from such bad practices. Regarding the quality of findings, it is shown in this chapter that there is a lack of evidence to suggest which commonly used methodology is most capable of producing reliable e-waste estimates.

While agreements are found over many parameters, disagreements are abundant too. Yet, it is also demonstrated in this chapter that agreements among studies do not necessarily mean that they are accurate. Conversely, seemingly outlying findings may be closer to the truth because of an appropriate distribution model is used. In all cases, despite the disappointing conclusion regarding the quality of e-waste studies for China, the lack of reliable data on domestic generation and disposal rate of e-waste in China should not prevent adoption of precautionary principles for e-waste management because there is already common consensus that hazardous e-waste recovery and disposal activities have been taking place at a significant scale in China. The sheer size of China also means that what China does or does not do substantially impacts the global environmental, for better or for worse.

As a result, there is no question about the need for institutionalizing proper e-waste control and making available environmentally sound e-waste recycling infrastructure. However, in view of the lack of information on the capacity needed, an incremental approach can be adopted to avoid excessive investment in this aspect.

Acknowledgments

This research is financially supported by the Public Policy Research Grant of the Research Grant Council of the Hong Kong Government (HKBU 2001-PPR-5), the Marginally-Funded Post-Graduate Fund (on Persistent Toxic Substances) from the Research Grants Council of Hong Kong, the Faculty Research Grant (FRG1/08-09/037) of the Hong

Kong Baptist University and the Hong Kong Polytechnic University research project, "Cross-Border Environmental Protection Between Hong Kong and Guangdong Province" (a/c: 1-BB05). (5938 words).

REFERENCES

Aizawa, H.; Yoshida, H., & Sakai, S.I. (2008). Current results and future perspectives for Japanese recycling of home electrical appliances. Resources, *Conservation & Recycling,* 52, 1399-1410.

Asia-Pacific Regional Centre for Hazardous Waste Management Training and Technology Transfer. (2005). Report on the Survey of the Import and the Environmentally Sound Management of Electronic Wastes in the Asia-Pacific Region. Beijing, China: Basel Convention Regional Centre in China, Tsinghua University.

Californian Integrated Waste Management Board. (2001). Selected E-waste Diversion in California: A Baseline Study. Sacramento, CA. November, #610-01-008.

China Business Council for Sustainable Development. Home Appliance Makers to Pay for Waste Recycling. 2007 [cited 2008 6 23) Available from: URL: http://english.cbcsd.org.cn/dynamic/associator/3177.shtml

China Statistics Bureau. China statistical yearbook 2009. Beijing: China Statistical Press; 2009. (in Chinese)

Chung, S.S., 2010. Projecting municipal solid waste: the case of Hong Kong SAR. Resources, *Conservation & Recycling,* 54, 759-768.

Chung, S.S., Lau, W. K.Y., & Zhang, C. (2011). The generation of and control measures for e-waste in Hong Kong. *Waste Management,* 31, 544-554.

Cobbing, M. Toxic Tech: Not in our backyard. Uncovering the Hidden Flows of E-waste. Amsterdam: *Greenpeace International;* 2008.

Crowe, M., Elser, A., Göpfert, B., Mertins, L., Meyer, T., Schmid, J., Spillner, A., & Ströbel, R. Waste from Electrical and Electronic Equipment (WEEE) – Quantities, Dangerous Substances and Treatment Methods. Copenhagen: *European Environment Agency;* 2003.

Duan, H.B., Yang, L.W., Eguster, M., Streicher-Porte, M., & Li, J.H. (2007). Waste electrical and electronic equipment generation: a China perspective. In *Proceedings of the Second International Conference on Waste Management and Technology*. (18-19 July, pp. 255-263). Beijing, China.

Environment Bureau. Safe and Sustainable: A New Producer Responsibility Scheme for Waste Electrical and Electronic Equipment. Hong Kong: Hong Kong SAR Government; 2010.

Feszty, K., Murchison, C., Baird, J., & Jamnejad, G. (2003). Assessment of the quantities of waste electrical and electronic equipment (WEEE) in Scotland. *Waste Management & Research*, 21, 207-217.

Florence, Z., & Price, T.J. (2005). Domestic-fridge recycling in Wales: mountains or mole hills? *Applied Energy*, 80, 125-140.

Greenpeace. Recycling of electronic waste in China and India: workplace and environmental contamination. August, Greenpeace International; 2005.

Hsu, E., & Kuo, C.M. (2005). Recycling rates of waste home appliances in Taiwan. *Waste Management,* 25, 53-65.

Kellow, A. (1999). Baptists and bootleggers? The Basel Convention and metals recycling trade. *Agenda,* 6, 29-38.

Leung, A.O.W., Luksemburg, W.J., Wong, A.S., & Wong, M.H. (2007). Spatial distribution of polybrominated diphenyl ethers and polychlorinated dibenzo-p-dioxins and dibenzofurans in soil and combusted residue at Guiyu, an electronic waste recycling site in southeast China. *Environmental Science & Technology,* 41, 2730-2737.

Li, J.H., & Duan, H.B. Study on best available technology of WEEE Treatment for less-developed countries. In*: Proceedings on International Conference on Waste Engineering and Management: ICWEM 2008*. 28-30 May. Hong Kong, 2008; 38-51.

Liu, X.B., Tanaka, M., & Matsui, Y. (2006a). Electrical and electronic waste management in China: progress and the barriers to overcome. *Waste Management & Research,* 24, 92-101.

Liu, X.B., Tanaka, M., & Matsui, Y. (2006b). Generation amount prediction and material flow analysis of electronic waste: a case study in Beijing, China. *Waste Management & Research,* 24, 434-445.

Lohse, J., Winteler, S., & Wulf-Schnabel, J. Final Report: Collection Targets for Waste from Electrical and Electronic Equipment (WEEE). Report compiled for the Directorate General (DG XI) Environment. May. Hamburg: Nuclear Safety and Civil Protection of the Commission of the European Communities; 1998.

Matthews, H.S., McMichael, F.C., Hendrickson, C.T., & Hart, D.J. Disposition and End-of-Life Options for Personal Computers. Pittsburgh: Carnegie Mellon University; 1997.

Müller, E., Schluep, M., Widmer, R., Gottschalk, F., & Böni, H. Assessment of E-Waste Flows: A Probabilistic Approach to Quantify E-waste Based on World ICT and Development Indicators. In: Proceedings of R'09 World Congress. 14 September, Davos; 2009.

Nnorom, I.C., & Osibanjo, O. (2008). Electronic waste (e-waste): material flows and management practices in Nigeria. *Waste Management,* 28, 1289-1498.

Puckett, J., & Byster, L. Exporting Harm: the High-Tech Trashing of Asia. 2002 [cited 2010 6 25]. Available from: URL: http://www.ban.org/E-waste/technotrashfinalcomp.pdf

Schluep, M., Hagelueken, C., Kuehr, R., Magalini, F., Maurer, C., Meskers, C., Mueller, E., & Wang, F. Recycling – from E-waste to Resources. July. Kenya: UNEP and United Nations University; 2009.

Steubing, B., Böni, H., Schluep, M., Silva, U., & Ludwig, C. (2010). Assessing computer waste generation in Chile using material flow analysis. *Waste Management*, 30, 473-482.

Streicher-Porte, M., & Yang, J. WEEE recycling in China: Present situation and main obstacles for improvement. In: *The IEEE International Symposium on Electronics and the Environment (ISEE).* 7-10 May, Orlando; 2007.

Walk, W. Approaches to estimate future quantities of waste electrical and electronic equipment (WEEE). In: *Proceedings of the Electronics Goes Green 2004*. 6-8 September, Berlin; 2004; 263-268.

Walk, W. (2009). Forecasting quantities of disused household CRT appliances – A regional case study approach and its application to Baden-Württemberg. *Waste Management,* 29, 945-951.

Wang, Y.C., Ru, Y.H., Veenstra, A., Wang, R.J., & Wang, Y. (2010). Recent developments in waste electrical and electronics equipment legislation in China. *International Journal of Advanced Manufacturing and Technology,* 47, 437-448.

Wilkinson, S., Duffy, N., & Crowe, M. Waste from electrical and electronic equipment in Ireland: a status report. May, Ireland: Environmental *Protection Agency;* 2001.

Wong, C.S.C., Duzgoren-Aydin, N.S., Aydin, A., & Wong, M.H. (2007). Evidence of excessive releases of metals from primitive e-waste processing in Guiyu, China. *Environmental Pollution,* 148, 62-72.

US Environmental Protection Agency. Municipal solid waste in the United States: 2000 Facts and Figures. June, EPA530-R-02-001, Office of Solid *Waste and Emergency Response,* USA; 2002.

Xinhua News Agency of China. China Suffers from Hazardous Waste. 2006. [cited 2011 1 2]. Available from: URL: www.china.org.cn/english/environment/192237.htm.

Yang, J.X., Lu, B., & Xu, C. (2008). WEEE flow and mitigating measures in China. *Waste Management,* 28, 1589-1597.

Yang, Y., & Williams, E. (2009). Logistic model-based forecast of sales and generation of obsolete computers in the US. *Technological Forecasting and Social Change,* 76, 1105-1114.

Yoshida, A., Tasaki, T., & Terazono, A., 2009. Material flow analysis of used personal computers in Japan. *Waste Management,* 29, 1602-1614.

Yu, J.L., Williams, E., Ju, M.T., & Yang, Y., 2010. Forecasting Global Generation of obsolete personal computers. *Environmental Science & Technology,* 44, 3232-7.

In: E-Waste: Management, Types and Challenges
Editors: Yuan Chun Li and Banci Lian Wang
ISBN 978-1-61942-217-9

Chapter 5

E-WASTE ENVIRONMENTAL CONTAMINATION AND PUBLIC HEALTH EFFECTS IN GUIYU, SOUTHEAST CHINA

Hao Ban, Xijin Xu, Kusheng Wu, Di Sun, Jian Zhang, Okunola Adenrele Alabi and Xia Huo

Analytic Cytology Laboratory and the Key Immunopathology Laboratory of Guangdong Province, Shantou University Medical College, Guangdong, China

ABSTRACT

Electronic waste (e-waste), which contains hazardous material, is much more expensive and difficult to dispose and recycle than other kinds of common garbage. E-waste has raised many concerns worldwidely due to its rapid generation; trans-boundary shipment from developed countries; improper recycling in developing countries; environmental impacts and health hazards. Property tags showed that US, Japan and UK are major exporters of these wastes to Guiyu, a town 40 km from the Shantou, Guangdong province, China. Local environment matrix, such as air, soil, water, sediment and leachate, has been found containing high concentration of hazardous material routed from e-waste recycling process. With extremely primitive and uncontrolled methods, nearly 6,000 family workshops in Guiyu process approximately 1.6 million tons of e-waste yearly and produce extensive pollutants which have adverse effects threatening local community. Majority of these workers are also residents of these recycling sites thereby increasing their exposure rates. Our studies found that people employed in e-waste recycling in Guiyu had a high incidence of skin damage, headache, vertigo, nausea, chronic gastritis, and gastroduodenal ulcers. Guiyu children and neonates have significantly elevated blood lead, cadmium and chromium levels with impairment of neurobehavioral development and damaged lymphocyte DNA. Guiyu neonates also showed higher rates of adverse birth outcomes such as preterm delivery, stillbirths and low birth weight. Our studies suggest that environmental pollution by uncontrolled e-waste process has adversely affected health condition of local population, especially fetus or children health and development, which may cause long-term impairments for them.

INTRODUCTION

Electronic waste, or e-waste is a waste type consisting of any broken, obsolete or unwanted electrical or electronic device or tools such as computers and computer-related equipment (keyboard, monitors, printers, etc), mobile phones, televisions, radios, VCRs, PDAs, photocopiers, DVD players, cameras, electric cables, LCD/plasma screens, microwave ovens for household, offices, schools, entertainment and any other consumer and business used, which are at the end of their useful life. Over the past ten years, consumer's demand for electronic items has increased rapidly. Obsolete electronic devices are increasing at an astonishing rate and have become the world's largest source of waste. For example, e-waste constitutes from 2% to 5% and 8% of all municipal solid waste streams in US and EU, increasing rapidly at around 4% per year or 3 times greater than normal municipal solid waste [1-4]. Every year, 20 to 50 million tones of e-waste are generated worldwidely, enough to load a train that could stretch around the world [5]. More than 500 million computers became obsolete in the US alone between 1997 and 2006 [6]. Because of the far more unsaturated global computer market and the rapidly decreasing average lifespan of a PC: 4-6 years in 1997 to only 2 years in 2005 and further decreasing [7], PC obsolescence now exceeds production. Other household or personal IT appliances also become outdated while being replaced rapidly due to their swift generation. For example, average working life of a mobile phone is 7 years, but worldwidely, an average consumer changes the mobile phone every 11 months [8]. In 2006, it was estimated that each year 130 million and 105 million mobile phones in the US and Europe will be thrown away, respectively [9]. In US, consumers had to switch from analog to digital TV service when cable companies stopped broadcasting analog signals on Feb. 17th, 2008, thereby resulting in mass junks of analog TVs. In China, about 20 million electronic household appliances reach the end-of-life every year, including TV, washing machines, PCs etc and 70 million cell phones [10]. It is reported that 400 million units of e-waste per year will be scrapped during the rest of this decade over the world [11]. Hence e-waste is one of the fastest growing waste types in the world.

One UN study has found that manufacturing a computer and its screen takes at least 240kg of fossil fuel, 22 kg of chemicals and 1.5 tones of water, more than the weight of a car [5]. An average PC weighs 29.6 kg and consists of metal (43.7%), plastics (23.3%), electronic components (17.3%) and glass (15%) [12]. As a component part for reuse, e-waste becomes a source for secondary raw material: 1 ton of circuit boards contains 130 kg copper, 0.45 kg gold, 20 kg tin, 2 kg silver and other metals such as cadmium, aluminum, lead, zinc and platinum, as well as glass and plastic. E-wastes contain both valuable materials as well as hazardous component, which require special handling and recycling methods.

HAZARDS OF E-WASTE

E-waste contains more than 1000 different substances, many of which are toxicants ranging from heavy metals to persistent organic pollutants (POPs). When e-wastes are improperly dismantled and recycled, toxic metals (e.g. lead, mercury, arsenic, cadmium, selenium, hexavalent chromium, barium, nickel, cobalt, gold, silver, copper) and POPs, like dioxins, furans, polybrominated diphenyl ethers (PBDEs), polychlorinated biphenyls (PCBs), polyvinyl chloride (PVC) and polycyclic aromatic hydrocarbons (PAHs), will be released into the environment and cause severe damage to human health. Reports have shown that about 700 million obsolete phones discarded in 2005 contained an estimated 560,000kg of lead in the form of solder [13]; 22% of the yearly world consumption of mercury was used in electronics manufacture [14] and 70% of the heavy metals (mercury and cadmium) in the US landfills came from electronic waste and 40% lead in landfills came from electrical and electronic equipment [15].

The e-waste harm to the environment and human health is much more severe than that of any other common wastes because e-waste not only contains a range of hazardous substances but is also extremely difficult to recycle and dispose. For example, a button battery can make 600 tons of water undrinkable, amount enough for a person's life consumption.

TRANSPORTATION OF E-WASTE

Because of the complicated construction and non-uniformity material content of e-waste, large scale recycling with mechanical automation is almost impossible. In some developed countries, due to the high expenditure of e-waste manual handling and increasing cost of e-waste recycling prescribed by strict environmental legislation, some e-waste disposal enterprises illegally export their collection to developing countries, where workers recycle e-waste using primitive technique with little regard of environmental protection for their safety. According to some data, more than 80% of the United States' e-waste was exported to Asia and Africa, such as India, China, Pakistan and Nigeria, 90% of which ends up in China. Guiyu, a town located about 250 km northeast of Hong Kong, China, is one of the prime destinations for such e-waste [16-19]. Property tags further show that Japan and UK, aside US, are contributors to the e-waste in China.

In 2005, US generated 2.6 million tons of e-waste, out of which only 12.6% was recycled [20]. In 2008, several reports showed the recycling industry in US was exploding, but some so-called recyclers were shipping the waste overseas [21]. In August 2008, a report by the US Government Accountability Office showed that 1,200 (mostly tiny) companies generated revenue of more than $3 billion in 2007 in US. The same report by them included a sting investigation which found that 43 US recycling firms, involving a large electronics recycler "Supreme" in New Jersey, were willing to sell e-waste, including broken monitors with cathode-ray tubes to buyers in foreign countries without getting the required permission from the Environmental Protection Agency (EPA) and the receiving nations [22]. CBS *60 Minutes* also found that e-waste recyclers were shipping the waste overseas [23].

Since the early 1990s, an international treaty known as the Basel Convention has restricted trade in hazardous waste, but the US is the only industrialized country that failed to ratify the pact. As one limited response to the Basel initiative, the EPA adopted civil rules taking into effect in January 2007, forbidding US companies from exporting monitors and televisions with cathode-ray tubes unless they have approval from the EPA and the receiving country.

E-Waste in Guiyu

China has been accepting vast quantities of discarded televisions, computers, printers, and other equipment from abroad since the early 1990s. Guiyu in China is one of the destinations for such e-waste [21-23]. Guiyu, a town with 27 villages in Chaoyang district, Shantou City, Guangdong Province, China (Figure 1), is one of the biggest and most notorious e-waste processing sites in the world. With a total area of 5, 200 hectares and local population of 139,000, Guiyu has accommodated millions of tons of e-waste from overseas and China domestic each year.

Because of low lying and frequently waterlogged farmlands, Guiyu residents were supplementing their farm income by the trash business before 1950s. Then the collection of plastic, glass, poultry feather, livestock bone, copper, tin etc. was started for processing and sale.

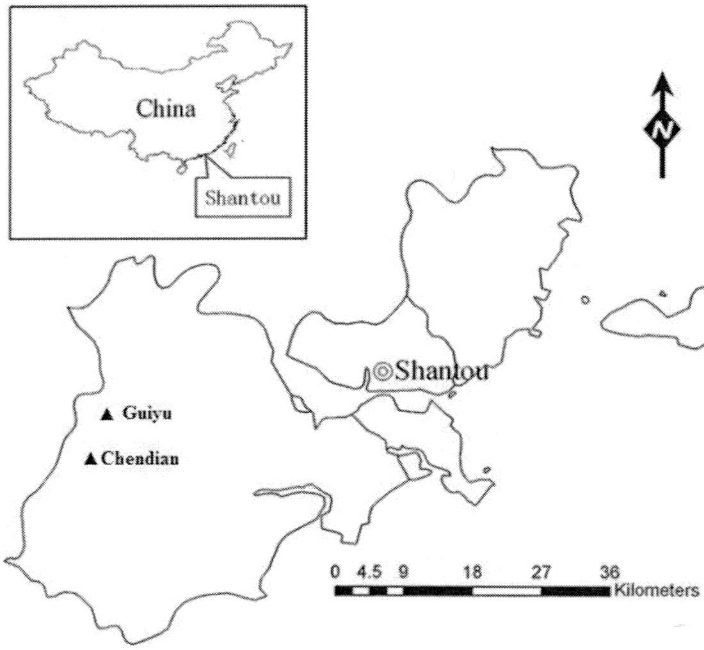

Figure 1.

It was a natural groove to follow when the electrical and electronic garbage output began to increase through the 1980s and 90s. Since then, the residents have been heavily involved in e-waste recycling activities.

Status of E-Waste Recycling in Guiyu

The bulk of recycling activity is carried out in 21 villages in Guiyu with over 5,500 individual family workshops, over 300 private companies and about 60,000 local residents.

Annually, Guiyu receives an estimated 1.55 million tons of e-waste, which is processed into 0.95 million tons of recycled plastic, 0.55 million tons of recycled metals and 0.05 million tons of reuse electronic components. It is estimated that 50,000 to 100,000 migrant laborers are employed in the trade. It is also noted that at least one worker of every family involves in the recycling work.

With methods of sorting, disassembling, smashing, burning and melting, saleable electronic components and raw materials are obtained in extremely primitive, archaic and uncontrolled process that promotes extensive environmental pollutants threatening local communities. Majority of the workers are also residents of this recycling site, which thereby increases their exposure level.

The primitive e-waste recycling procedure in Guiyu is mainly as follows: *a*) With simple tools such as electric drill, cutter, hammer and screwdriver, old electronic equipment is dismantled into component parts (Figure 2), including monitor, hard drive, CD driver, wires, cables, circuit boards, transformer, charger, battery, and plastic or metal frame that are sold for reuse or further recycling by other workshops. *b*) Circuit boards (Figure 3) of computers and other large appliances are heated over coal fires to melt the solder and release valuable electronic components, such as diodes, resistors, and microchips. *c*) Circuit boards of cell phones, pager and other hand-held devices are heated for the solders by an electro-thermal machine (Figure 4). *d*) Memory banks, chips or circuit boards, which contain tiny amounts of gold and palladium, are treated in acid bath with a medieval acid recipe - aqua regia, a highly corrosive and dangerous fuming yellow or red solution to extract precious gold and palladium,

during which workers just stand beside the acid bath without enough protective measures. The e-waste parts are washed with the solution frequently.

Figure 2.

Figure 3.

Then the waste acids are discharged into nearby fields and stream. *e*) Wires and cables are stripped or simply burnt in open air to recover metals (Figure 5). *f*) Transformers, chargers and other plastic or rubber coating components are burnt in a coal fire.

After burning, the workers can take out the copper or other metals easily with hammer and fingers (Figure 6). *g*) Printer cartridges are swept for printer toner (Figure 7). *h*) Cathode ray tubes were just simply hammered open for metals such as lead, mercury, cadmium, phosphors and barium. *i*) Plastics [e.g., polyvinyl chloride (PVC), acrylonitrile butadiene styrene copolymer (ABS) and high-density polyethylene (HDPE)] are sorted by workers according to the rigidity, color, and luster. Plastic scraps that can not be sorted visually are burnt and classified by burning odor. "When you burn it, you can tell what kind of plastic it is.

Figure 4.

Figure 5.

They smell different. There are many kinds of plastic, probably 60 or 70 types." said one of the workers who smell and distinguish plastic. Another way to sort different plastics granule is gravitational separation into ceramic jugs with brine (Figure 8), after which, the granule are exposed on the sidewalk to dry. There are no safety assessment and protective equipments for handling e-waste in all of the above procedures. Although the methods for processing e-waste are primitive, the coordination of e-waste recycling in Guiyu is very well organized into specific tasks: five of Guiyu villages specialize mainly in old equipments dismantling; six in circuit-board disassembly; two in wire and cable disassembly and seven in plastics and metals reprocessing. The chain of recycling components from each type of e-waste is well established in Guiyu town.

Figure 6.

Figure 7.

Figure 8.

ENVIRONMENTAL CONTAMINATION OF GUIYU

1. Persistent Organic Pollutants (POPs) in the Environment of Guiyu

Open, uncontrolled burning of chlorinated and brominated plastics is known worldwidely to cause the emission of polychlorinated and polybrominated dioxins. These are among the most toxic compounds known on earth. The roster of substances liberated during e-waste processing is a toxicological nightmare. Known carcinogens like dioxins and polycyclic aromatic hydrocarbons, neurotoxic elements like lead and brominated flame retardants including PBDEs, which have been shown to disrupt endocrine system in laboratory animals and wildlife are constantly released into the environment. Dr. Eddy Zeng from Guangzhou Institute of Geochemistry reported that about 76,000 metric tons of PBDEs alone are released into the environment each year at e-waste sites in China [24]. More and more of these toxic chemicals are getting into the food supply.

1.1. Polybrominated Diphenyl Ethers (PBDEs)

The Hong Kong team, led by Minghong Wong, had undertaken pioneering work to track the fate of e-waste toxicants. Along a riverbank in Guiyu, for example, they found levels of PBDEs were thousands of times higher than those found in soil from a control site in the province. Soil from a site where acid leaching was used to recover valuable metals, contained up to 4250 ng/g PBDEs [25]. Luo et al. [26] reported PBDEs concentrations of 191 to 9156 ng/g (dry weight) in farmland soil 2 km from an e-waste recycling workshop. In all samples, BDE-47, BDE-100, BDE-99, BDE-154, BDE-153 and BDE-183 were the dominant congeners. Total mono-to-hepta brominated diphenyl ether (BDE) homologue concentrations (1140 and 1169 μg/kg dry wt.) in soil near dumping sites were approximately 10-60 times higer than those reported for other PBDEs-contaminated locations in the world [27]. Another report compared PBDE congeners concentrations in air samples among Hong Kong, Guangzhou and Guiyu. All PBDE congeners tested in Guiyu were up to 58-691 times higher than other urban sites and more than 100 times higher than other studies reported elsewhere [28]. Another report compared diurnal variation of PBDEs in the atmosphere between Guiyu and Chendian, PBDEs concentrations in Guiyu were 11- to 79-fold higher than those in Chendian in the daytime and 11- to 37-fold at night, respectively [29]. Luo et al. [30] reported that carp from Nanyang River, near Guiyu, was bioaccumulating PBDEs to concentrations of up to 766 ng/g (fresh weight). Unsurprisingly, the same group reported elevated PBDEs concentrations in the sediments (up to 16,000 ng/g) of the same river [31]. Wu et al. [32] found 1091 ng/g PBDEs on a wet weight basis of the water snake in Guiyu. Total PBDEs concentration in open burning site was more than 7200 times greater than that at reservoir area. Of the PBDE homologues, deca-BDE (BDE-209) had the highest abundance (63-81%) at all the sites [17].

1.2. Polycyclic Aromatic Hydrocarbons (PAHs)

The concentrations of PAHs in soil obtained from a printer roller dump site was 593 μg/kg dry weight (dry wt.) and in sediment from a duck pond was 514 μg/kg (dry wt.) in the e-waste recycling site of Guiyu [33]. The concentrations of a sum of 16 PAHs ranged from 44.8 to 3206 μg/kg (dry wt.), in the descending order of e-waste open burning sites (2065 μg/kg) > areas near burning sites (851 μg/kg) > rice fields (354 μg/kg) > reservoir areas (125μg/kg). The dominant PAHs were naphthalene, phenanthrene and fluoranthene, which were mainly derived from incomplete combustion of e-waste (e.g. wire insulations and PVC materials) [34,35]. Among mass concentrations contained in $PM_{2.5}$ (particles with aerodynamic diameter smaller than 2.5 of μm, implicated in pulmonary disease and cardiovascular disease) in Guiyu, five-ring and six-ring PAHs accounted for 73% of total PAHs. The average concentration of benzo(a)pyrene was 2-6 times higher than in other Asian cities [3].

1.3. Polychlorinated Biphenyls (PCB)

Sediment from Lianjiang River of Guiyu was found to be contaminated by PCB (743 μg/kg) at a level approaching three times higher than the Canadian Environmental Quality

Guidelines probable effect level of 277 μg/kg [33]. The PCB concentrations of Guiyu air samples were 12-18 times higher than those in Chendian, 9 km away, and 37-133 times higher than samples taken in Guangzhou, a town located 450 km from Guiyu [36]. Wu et al. [32] reported average 16,512 ng/g PCB on a wet weight basis of the water snake in Guiyu.

1.4. Polychlorinated Dibenzo-P-Dioxin and Polychlorinated Dibenzofuran (PCDD/F)

PCDD/F concentrations in samples collected from the blackened e-waste and burning ash were as high as 469 and 5678 pg I-TEQ g^{-1}, respectively. PCDD/F concentrations in river sediments collected near this e-waste recycling site were 21-39,000 pg I-TEQ g^{-1} [37]. During the past decades, seven PCDD/Fs pollution hotspots (Ya-er Lake, Hubei Province, Chinese Schistosomiasis affected areas of Jiangxi Province, e-waste recycling area of Yangtze River Delta, e-waste recycling area of Guiyu, Pearl River Delta Kwun Tong, Victoria Harbour, Hong Kong, southern Taiwan) have been found, e-waste recycling area of Guiyu, with 967,500 pg/g in mixture of burnt residue and soil, which is the highest level among all hotspots. The total PCDD/F concentrations ranged from 466 to 599,156 pg/g [34,38].

2. Heavy Metal Pollution in Guiyu

The environment of Guiyu has been severely contaminated with a variety of toxic chemicals including heavy metals (e.g., Cd, Co, Cr, Cu, Ni, Pb, Zn). Not only are the rivers polluted, also the underground water, which is as a result of the dumping and filtering of toxic chemicals into the ground. The air quality was also drastically affected [39]. Soil of Guiyu was reported to contain heavy metals such as Cd, Cr, Cu, Ni, Pb and Zn which were higher than the threshold of New Dutch List [40]. Wong et al. [41] reported that dissolved metal concentrations were high in the two rivers within Guiyu, including Ar, Ag, Be, Cd, Cr, Cu, Li, Mo, Ni, Pb, Sb, Se and Zn. Temporal distributions of the metals suggested recent discharges of metals attributable to a strong acid leaching operation of e-waste. Pb isotopic compositions of dissolved Pb confirmed that more than one non-indigenous Pb were present in these two rivers. Wong et al. [42] also investigated river sediments in Guiyu, which were found contaminated with Cd, Cu, Ni, Pb and Zn compared to the uncontaminated sediments. Leung et al. [43] reported that Pb, Cu, Zn and Ni mean concentrations were elevated in workshop dust and in dust of adjacent roads in Guiyu. A comparative study of Asian countries [3] showed that concentrations of Cr, Cu and Zn in $PM_{2.5}$ of Guiyu were 4-33 times higher than in other Asian countries.

2.1. The River Pollution

Many rivers in Guiyu has been polluted by heavy metal, partially the increasing levels of Pb found in water sources [41,44]. The water and sediments in rivers has been found containing high levers of heavy metals [45]. It was reported that the concentrations of Cu, Pb, Sn, Ni, and Cd in sediments taken from discharging channels were between 400 and 600 times higher than what would be expected for uncontaminated river sediments [46]. Studies on river sediment in the two main rivers running through Guiyu, the Nanyang and the Liangjiang, have found elevated levels of Cu, Cd, Ni, Pb, and Zn. Similarly, water samples in the rivers have been found containing significantly elevated levels of Cd, Cu, Ni, and Ag [47].

There was also a study showing the Cu, Pb, and Ni concentrations of sediments collected from Guiyu and Lianjiang River were higher than those measured in river sediment in Shenzhen and Hong Kong [33,48]. In another study, surface water and sediment collected from Lianjiang River and its tributaries were analyzed for concentrations of Cu, Pb, Ni, Cd, Hg, and As. The results showed concentrations of Cu from surface water in the Guiyu area were 2.4 to 131 times the reference background concentration, followed in descending order by Ni, Cd, Pb, Hg, and As. Concentrations of Cu in sediment samples in Guiyu area were 3.2 to 429 times the reference background concentration, followed by Ni, Hg, Pb, Cd, and As [49].

2.2. The Air Pollution

The air was found containing particles loaded with a number of heavy metals [3], so the heavy metal pollution in the dust is usually tested to reflect the air pollution. In Guiyu, the concentration of Pb in dust collected from the workshops was hundreds of times higher than typical levels reported for indoor dust in other parts of world [46]. To evaluate the extent of heavy metals (Cd, Co, Cr, Cu, Ni, Pb, Zn) contamination from printed circuit board recycling, surface dust samples were collected from recycling workshops, adjacent roads, a schoolyard, and an outdoor food market. ICP-OES analyses revealed elevated mean concentrations in workshop dust (Pb 110 000 mg/kg, Cu 8360 mg/kg, Zn 4420 mg/kg, and Ni 1500 mg/kg) and in dust of adjacent roads (Pb 22 600 mg/kg, Cu 6170 mg/kg, Zn 2370 mg/kg, and Ni 304 mg/kg). Pb and Cu in road dust were 330 and 106, and 371 and 155 times higher, respectively, than non e-waste sites located 8 and 30 km away [43].

2.3. The Soil Pollution

It is worth noting that approximately 30% of the agricultural practice in this region solely relies on irrigation water from Lianjiang River. The continuous processing of e-waste recycling in Guiyu and elevated heavy metal concentrations in Lianjiang River may thus result in elevated metal loadings of soils and crops [50]. A study showed that in Guiyu, Cd, Sn, and Ba were found at levels 1388, 152 and 10 times respectively higher than the EPA threshold for environmental risk in the soil [51]. Soils contaminated by heavy metals from either aerial depositions or irrigation were likely to induce a corresponding contamination in harvested crops [52]. Crops in or close to contaminated sites could uptake and accumulate these metals, and then exerted potential risk to humans and animals [53,54].

All the primitive recycling processes emit chocking fume and toxic ash. Hence Guiyu was called one of the most toxic places on earth and compared to be a sort of Chernobyl of e-waste [16].

HAZARDOUS EFFECTS OF E-WASTE CONTAMINATION ON PUBLIC HEALTH IN GUIYU

Over the past few years, many published studies clearly indicate that there is a causal relationship between the pollution levels and emissions from the informal e-waste recycling [55]. Consequently, increasing epidemiological and clinical evidence has raised concern about the potential hazards towards local population.

There is a great exposed population in Guiyu because these people are not just working with these wastes, they are living with them. E-waste combustion generated small particles ($PM_{2.5}$) in Guiyu exceed the US EPA 24-h $PM_{2.5}$ ambient air quality standard, exposed to which, the pregnant women and children are more vulnerable [55]. Study by Bi et al. [56] showed that the median BDE-209 concentration of dismantling workers in Guiyu was 50-200 times higher than previously reported occupationally exposed population. The highest BDE-209 concentration was 3100 ng/g in lipid, the highest ever reported in humans.

Researchers have reported that PBDEs blood levels in Guiyu residents were on average, nearly 600ppb. These were the highest PBDEs levels ever reported to date in people over the world. The Guiyu levels were 10 times higher than average levels in the US and more than 100 times higher than in Europe [24].

Researchers found that the inhaled dioxin of adults in Guiyu was 68.9 and 126 pg TEQ/kg/day in the summer and winter, respectively, which was 17-31 times higher than the WHO recommended maximum dioxins inhalation [57]. However, the daily dioxin intake doses of Guiyu children were about 2 times higher than that of adults [34].

In 2003, our research team began to conduct health investigation on local public in Guiyu. We conducted health surveys for 638 residents and recycling workers. A total of 504 children and 423 neonates have been examined for health assessment and tested blood for heavy metals such as Pb, Cd and Cr. Evidence suggested that workers in Guiyu have high incidence of skin damage, headache, vertigo, nausea, chronic gastritis, gastric and duodenal ulcers [58]. These symptoms usually present in chronic lead poisoned individuals.

Fetus and children are more vulnerable to these hazardous substances, because heavy metals and POPs have the capacity to pass through the placental barrier and get into the fetal blood stream. Their body tissues, organs, and biological systems are developing at a rapid rate, but function of absorption, metabolism, and elimination of toxic chemicals are less efficient. Young children are exposed to chemicals throughout the pre and post natal periods. They eat more, drink more, and breathe more in proportion to their body weight. Their behavior can result more exposure to harmful chemicals. Just as children naturally absorb more nutrients from their diets, they absorb more toxicants than adults. Children's bodies are smaller, so the same toxicant level represents a much larger proportion of it relative to a child's body weight. All of these factors put young, developing systems at particular risk for health impairment.

For the purpose of children health checking, in 2005, we studied blood lead levels (BLLs) in a cluster sample of 165 children younger than 6-year-old who lived in Guiyu. Analyses showed that children living in Guiyu had significantly higher BLLs compared with control group. We found that the elevated rates of BLLs in Guiyu in 2005 and 2006 are 81.8% and 70.8% respectively [19,59], which exceeded the defined values by the US Centers for Disease Control and Prevention (CDC) that considered elevated BLLs as those $\geq$ 10μg/dL in children $\leq$ 6 years of age [60]. However, even the BLLs are $<$ 10μg/dL, it is still unsafe for children. Intelligent quotients, academic performance, cognitive functioning and neurobehavioral developments of children decreased greater with increasing BLLs in $<$ 10μg/dL group than $\geq$ 10μg/dL group [61]. We are conducting research on the relationship between lead exposure and intelligence, behavior and executive functioning of Guiyu children, aim to assess the extent of lead exposure to children's neurobehavioral development. We found location of child residence engagement in work related to e-waste was risk factor to child BLL, activity level, approach-withdrawal, adaptability, and mood. Child hand washing before food consumption was a protected factor. There were close relationships between BLLs elevation, temperament alteration and the e-waste recycling activities in Guiyu [62]. Elevated lead exposed children were also reported with neurobehavioral deficits such as attention deficit hyperactivity disorder (ADHD), decreased attention span, and non-persistence, even impulsive and aggressive behaviors [63,64]. Lead exposure may cause devastating effects on nervous development, we have investigated mechanisms for neurotoxicity of lead by determining the effect on apoptosis and expression of related genes, *XIAP* (X chromosome-linked inhibitor of apoptosis protein) and *Smac* (second mitochondrial activator of caspase) in hippocampus of developing rats. Data indicated that exposure to lead acetate caused a dose-dependent and significant increase of apoptosis in the hippocampus of developing rats through depressing the expression of the *XIAP* gene, which may be an important mechanism of lead neurotoxicity [65].

Lead can freely cross the placenta via diffusion, and may cause fetal abortion or fetal death in some extreme cases. So we hypothesized that fetus was the first stage exposed to lead in Guiyu neonates. We developed a method for determination of meconium (first stool of the newborn) lead level of newborn by graphite furnace atomic absorption spectrometry [66]. In our study, it was shown that the cord blood lead concentrations of neonates were in high levels, which accounted for impairment of neonate's neurobehavioral development [67,68]. In 2009, we tested the placenta concentrations of Pb, Cd, Cr, and Ni (PCPb, PCCd, PCCr, and PCNi) from Guiyu. It was found that women living in Guiyu had elevated PCPb compared with the control area. The average concentration of PCCd was 3-26 times higher than other reported. According to this investigation, neonates from Guiyu had a slightly lower birth weight and significantly lower rate of birth term when compared with the control group [69].

We also examined the blood cadmium levels of Guiyu children. Compared with results from studies conducted in some other cities in China, Guiyu children had higher blood cadmium levels [59]. In order to investigate Cd exposure levels in neonates, we also determined the placental cadmium, cord blood cadmium and the expression of placental metallothionein (MT), which plays a part in cell growth and differentiation; anti-oxidation; anti-ionizing radiation; and metabolism of trace element. Cd is an effective inducer of MT synthesis, which in turn protects host tissues against Cd damage. Data revealed that the placental cadmium and cord blood cadmium levels in Guiyu neonates were higher than those in other cities in China and other countries. Immunohistochemistry showed positive MT expression in placenta was higher for Guiyu than the control area. MT level was increased and related with corresponding Cd content in cord blood and placenta. The results supported the fact of Cd pollution in Guiyu [70].

We also tested umbilical cord blood chromium levels (UCBCLs) of 200 Guiyu neonates and examined lymphocyte DNA damage with Comet Assay. There were significant differences between the results in UCBCLs of neonates in Guiyu and the control group, the same with the results in terms of DNA damage. Higher levels of chromium in neonates were found correlating with their mothers' exposure to e-waste recycling [71].

With our studies above, there was conclusive evidence that high concentration of heavy metals such as Pb, Cr and Cd in neonates or children living in e-waste recycling areas in Guiyu was related to toxicants exposure caused by environmental pollution of the primitive e-waste recycling activities.

We also conducted research about local children's exposure to POPs. PBDEs are known as flame retardants which have no chemical bonds with other e-waste components. Therefore, it is easy for them to release from the e-waste to the environment. PBDEs are lipophilic, which result in their bioaccumulation in tissues and biomagnification in food chains [28]. Exposure to PBDEs may have endocrine-disrupting effects. Majority of PBDEs toxicological studies have focused on thyroid hormone disruption while a smaller number on the abnormal of the estrogen/androgen hormone system [72]. Some research showed that prenatal PBDEs exposure may also cause developmental neurotoxic effects to children [73]. We evaluated the exposure of Guiyu's neonates to PBDEs by testing umbilical cord blood and the health effects. It was shown that PBDEs levels significantly differed in neonates by normal birth and adverse birth outcomes including stillbirth, low birth weight, and premature delivery. The neonates from Guiyu are exposed to high levels of PBDEs [74].

We also assessed the body burden of neonates of PCBs and potential health risks for neonates. Some individual PCB congeners tested in umbilical cord blood were found to correlate with reduced neonatal height, weight, Apgar scores, gestational age, and body mass index (BMI). Difference of total PCBs level was found between normal birth and adverse birth outcomes. Neonates from Guiyu are exposed to high levels of PCBs, and prenatal exposure to PCBs appears to influence neonatal health [75].

Generally speaking, our studies found that Guiyu children and newborns had significantly elevated blood lead, cadmium, and chromium levels, also found are PBDEs and PCBs in newborn cord blood. High PBDEs levels were reported in stillbirth, premature and low birth weight newborns, while PCBs have adverse effects to neonates' development.

Further investigations and research is necessary to reveal disease-related mechanism of e-waste hazards and show the correlation between e-waste contamination and public health effects.

The toxicity of e-waste not only impairs health of Guiyu children but also threatens US children. The US has ordered recalls of cheap toy, trinkets and jewelry from China because most of the pieces contained toxic lead. Some studies indicated that the toxic lead in the products comes from recovered materials of e-waste shipped from the US [4]. Chen et al. [76] reported that brominated flame retardants, including PBDEs, were found elevated in children's toys purchased from South China. Researchers pointed out that it is conceivable the manufacture of exported toys products use recycled materials from e-waste, which may contain brominated flame retardants, like PBDEs and PCBs [18].

CONCLUSION

As for the e-waste stream, unless there is a proper way to dispose the e-waste, shipping to China or other developing countries does not constitute a responsible approach for the developed countries who are supposed to preserve world environment and health. Besides advanced technology, consumer participation (e.g. collecting, sorting and dismantling properly) and official policy are necessary for the recovery of precious materials from e-waste.

On the other hand, advanced processing methods, proper protection, active environment measuring, and continuous health monitoring should be introduced and put in place for the exposed population. Our modern society must now find ways of safely and economically managing these wastes, which requires significant investment by governments, industry and individuals in technology and education. This is the only way we can preserve the global environment and make it safe for generations yet-unborn.

REFERENCES

[1] Kang, H. Y. and Schoenung, J. M. (2005). Electronic waste recycling: A review of US infrastructure and technology options. *Resources Conservation and Recycling, 4,* 368-400.

[2] Streicher-Porte, M. (2006). *Proceedings of the 2006 IEEE International Symposium on Electronics and the Environment: SWICO/S.EN.S, the Swiss WEEE recycling systems, and best practices from other European systems.* San Francisco, IEEE.

[3] Deng, W. J. and Louie, P. K. (2006). Atmospheric levels and cytotoxicity of PAHs and heavy metals in TSP and PM2.5 at an electronic waste recycling site in southeast China. *Atmospheric Environment, 40,* 6945-6955.

[4] Darby, L. and Obara, L. (2005). Household recycling behaviour and attitudes towards the disposal of small electrical and electronic equipment. *Resources Conservation and Recycling, 1,* 17-35.

[5] Schwarzer, S. and De Ono, A. E-waste, the hidden side of IT equipment's manufacturing and use. UNEP Early Warning on Emerging Environmental Threats. 2005. http://www.grid.unep.ch/product/publication/download/ew_ewaste.en.pdf.

[6] Yu, J. Q. and Welford, R. (2006). Industry responses to EU WEEE and ROHS Directives: perspectives from China. *Corporate Social Responsibility and Environmental Management, 5,* 286-299.

[7] Kang, H. Y. and Schoenung, J. M. (2006). *Proceedings of the 2006 IEEE International Symposium on Electronics and the Environment: End-of-life personal computer systems in California: analysis of emissions and infrastructure needed to recycle in the future.* San Francisco, IEEE.

[8] Sharpe, M. (2005). Climbing the e-waste mountain. *Journal of Environmental Monitoring, 10,* 933-936.

[9] Canning, L. (2006). Rethinking market connections: mobile phone recovery, reuse and recycling in the UK. *Journal of Business and Industrial Marketing, 4-5,* 320-329.

[10] Zhong, Y. G. and Yu, Q. D. (2006). *2006 IEEE International Conference on Service Operations and Logistics, and Informatics: A research into the recycling system of waste electrical and electronic equipment in China.* San Francisco, IEEE.

[11] International Association of Electronics Recyclers. (2006). *Electronics Recycling Industry Report.* Portland: Resource Recycling, Inc.

[12] Berkhout, F. and Hertin, J. (2004). De-materialising and re-materialising: digital technologies and the environment. *Futures, 8,* 903-920.

[13] Lincoln, J. D. and Ogunseitan, O. A. (2005). *Advanced Packaging Materials: Processes, Properties and Interfaces: Environmentally benign materials for electronics: a review of current developments and emerging technologies. Proceedings International Symposium on.* Portland: IEEE.

[14] Realff, M. J. and Raymond, M. (2004). E-waste: An opportunity. *Materials Today, 1,* 40-45.

[15] Widmer, R. and Oswald-Krapf, H. (2005). Global perspectives on e-waste. *Environmental Impact Assessment Review, 5,* 436-458.

[16] Halluite, J. and Linton, J. D. (2005). The challenge of hazardous waste management in a sustainable environment: insights from electronic recovery laws. *Corp. Soc. Responsib. Environ Manage, 12,* 31-37.

[17] Wong, M. H. and Wu, S. C. (2007). Export of toxic chemicals - a review of the case of uncontrolled electronic-waste recycling. *Environmental Pollution, 149,* 131-140.

[18] Robinson, B. H. (2009). E-waste: An assessment of global production and environmental impacts. *Science of the Total Environment, 408,* 183-191.

[19] Huo, X. and Peng, L. (2007). Elevated blood lead levels of children in Guiyu, an electronic waste recycling town in China. *Environ. Health Perspect., 115,* 1113-1117.

[20] United States Environmental Protection Agency. Municipal Solid Waste in the United States: 2005 Facts and Figures: 164. (2006). http://www.epa.gov/osw/nonhaz/municipal/msw99.htm.

[21] Walsh B. (2009). E-Waste Not. *Time, 2,* 49-50.

[22] Elgin, B. E-Waste: *The Dirty Secret of Recycling Electronics.* (2008). http://www.businessweek.com/magazine/content/08_43/b4105000160974.htm.

[23] CBS. (2008). Following The Trail Of Toxic E-Waste. *CBS news 60 Minutes.*

[24] Richard, S. (2009). CHINA Confronting a Toxic Blowback from the Electronics Trade. *Science, 325,* 1055.

[25] Leung, A. O. and William, J. L. (2007). Spatial distribution of polybrominated diphenyl ethers and polychlorinated dibenzo-p-dioxins and dibenzofurans in soil and combusted residue at Guiyu, an electronic waste recycling site in southeast China. *Environ. Sci. Technol., 41,* 2730-2737.

[26] Luo, Y. and Luo, X. J. (2009). Polybrominated diphenyl ethers in road and farmland soils from an e-waste recycling region in Southern China: Concentrations, source profiles, and potential dispersion and deposition. *Sci. Total. Environ., 407,* 1105-1113.

[27] Wang, D. L. and Cai, Z. W. (2005). Determination of polybrominated diphenyl ethers in soil and sediment from an electronic waste recycling facility. *Chemosphere, 60,* 810–816.

[28] Deng, W. J. and Zheng, J. S. (2007). Distribution of PBDEs in air particles from an electronic waste recycling site compared with Guangzhou and Hong Kong, South China. *Environ. Int., 33,* 1063-1069.

[29] Chen, D. H. and Bi, X. H. (2009). Pollution characterization and diurnal variation of PBDEs in the atmosphere of an E-waste dismantling region. *Environmental Pollution, 157,* 1051-1057.

[30] Luo, Q. and Wong, M. H. (2007). Determination of polybrominated diphenyl ethers in freshwater fishes from a river polluted by e-wastes. *Talanta, 72,* 1644-1649.

[31] Luo, X. J. and Zhang, X. L. (2009). Persistent halogenated compounds in waterbirds from an e-waste recycling region in South China. *Environ. Sci. Technol., 43,* 306-311.

[32] Wu, J. P. and Luo, X. J. (2008). Bioaccumulation of polybrominated diphenyl ethers (PBDEs) and polychlorinated biphenyls (PCBs) in wild aquatic species from an electronic waste (e-waste) recycling site in South China. *Environ. Int., 34,* 1109-1113.

[33] Leung, A. and Zong, W. C. (2006). Environmental contamination from electronic waste recycling at Guiyu, southeast China. *J. Mater. Cycles Waste Manag., 8,* 21-33.

[34] Yu, X. Z. and Gao, Y. (2006). Distribution of polycyclic aromatic hydrocarbons in soils at Guiyu area of China, affected by recycling of electronic waste using primitive technologies. *Chemosphere, 65,* 1500-1509.

[35] Guo, Y. and Wu, K. (2011). Sources, distribution and toxicity of Polycyclic aromatic hydrocarbons. *Journal of Environmental health, 73,* 22-25.

[36] Li, H. and Yu, L. (2007). Severe PCDD/F and PBDD/F pollution in air around an electronic waste dismantling area in China. *Environ. Sci. Technol., 41,* 5641-5646.

[37] Luksemburg, W. J. and Mitzel, R. S. (2002). Polychlorinated dibenzodioxins and dibenzofurans (PCDDs/PCDFs) levels in environmental and human hair samples around an electronic waste processing site in Guiyu, Guangdong Province, China. *Organohalogen Compd., 55,* 347–349.

[38] Zheng, G. J. and Leung, A. O. (2008). Polychlorinated dibenzo-p-dioxins and dibenzofurans pollution in China: Sources, environmental levels and potential human health impacts. *Environment International, 34,* 1050-1061.

[39] BAN (The Basel Action Network SVTC). Exporting harm: the high-tech trashing of asia. 2002. http://www.banorg/E-waste/technotrashfinalcomp.pdf.

[40] VROM. (2001). The New Dutch List. Intervention Values and Target Values: Soil Quality Standards. Netherlands Ministry of Housing, *Spatial Planning and Environment*, Department of Soil Protection, The Hague, Netherlands.

[41] Wong, C. S. and Duzgoren-Aydin, N. S. (2007). Evidence of excessive releases of metals from primitive e-waste processing in Guiyu, China. *Environmental Pollution, 148,* 62-72.

[42] Wong, C. S. and Wu, S. C. (2007). Trace metal contamination of sediments in an e-waste processing village in China. *Environmental Pollution, 145,* 434-442.

[43] Leung, A. O. and Duzgoren-Aydin, N. S. (2008). Heavy metals concentrations of surface dust from e-waste recycling and its human health implications in southeast China. *Environmental Science and Technology, 42,* 2674-2680.

[44] Wang, J. P. and Guo, X. K. (2006). Impact of electronic wastes recycling on environmental quality. *Biomed. Environ. Sci., 19,* 137-142.

[45] Walters, A. and Santillo, D. Evidence of environmental and health impacts of electronics recycling in China: an update. 2008. http://www.greenpeace.to/publications/impacts-of-e-recycling-China-update.pdf.

[46] Brigden, K. and Labunska, I. Recycling of electronic wastes in China and India: *work place and environmental contamination.* 2005. http://www.greenpeace.org/eastasia/publications/reports/toxics/2005/recycling-of-electronic-wastes/.

[47] Jill, H. and Shu, T. (2008). Environment and Health in the Twenty-First Century Challenges and Solutions. *Ann. N. Y. Acad. Sci, 1140,* 1–21.

[48] Cheung, K.C. and Poon, B. H. (2003). Assessment of metal and nutrient concentrations in river water and sediment collected from the cities in the Pearl River Delta, South China. *Chemosphere, 52,* 1431–1440.

[49] Guo, Y. and Huang, C. J. (2009). Heavy Metal Contamination from Electronic Waste Recycling at Guiyu, Southeastern China. *J. Environ. Qual, 38,* 1617-1626.

[50] Mueller, M. (1994). Distribution of cadmium in the food chain (soil-plant-human) of a cadmium exposed area and the health risks of the general population. *Sci. Total Environ, 156,* 151-158.

[51] Osibanjo, O. and Nnorom, I. C. (2007). The challenge of electronic waste (e-waste) management in developing countries. *Waste Management Research, 25,* 489-501.

[52] Nan, Z. and Zhao, C. (2002). Relations between soil properties and elected heavy metal concentrations in spring wheat (Triticum aestivum L.) grown in contaminated soils. *Water, Air, Soil Pollut, 133,* 205-213.

[53] Gupta, U. and Gupta, S. (1998). Trace element toxicity relationships to crop production and livestock and human health: implications for management. *Commun. Soil. Sci. Plant Anal, 29,* 1491-1522.

[54] Jarup, L. (2003). Hazards of heavy metal contamination. *Brit. Med Bull, 68,* 167-182.

[55] Alejandra, S. and Mathias, S. (2010). A review of the environmental fate and effects of hazardous substances released from electrical and electronic equipments during recycling: Examples from China. *Environmental Impact Assessment Review, 30,* 28-41.

[56] Bi, X. and Thomas, G. O. (2007). Exposure of electronics dismantling workers to polybrominated diphenyl ethers, polychlorinated biphenyls, and organochlorine pesticides in South China. *Environ. Sci. Technol., 41,* 5647-5653.

[57] Chatterjee R. (2007). E-waste recycling spews dioxins into the air. *Environ. Sci. Technol., 41,* 5577-5577.

[58] Qiu, B. and Peng, L. (2004). Medical investigation of e-waste demanufacturing industry in Guiyu town. In: Proceedings of the International *Conference on Electronic Waste and Extended Producer Responsibility*, 79-83.

[59] Zheng, L. and Wu, K. (2008). Blood lead and cadmium levels and relevant factors among children from an e-waste recycling town in China. *Environ. Res., 108,* 15-20.

[60] CDC (Centers for Disease Control). Preventing Lead Poisoning in Young Children. Atlanta, GA: *Centers for Disease Control.* 1991. http://wonder.cdc.gov/wonder/prevguid/p0000029/p0000029.asp#head001000000000000.

[61] Wang, Q. and Zhao, H. H. (2009). Adverse health effects of lead exposure on children and exploration to internal lead indicator. *Science of the Total Environment, 407,* 5986-5992.

[62] Liu, J. and Xu, X. (2011). Association between lead exposure from electronic waste recycling and child temperament alterations. *Neurotoxicology, 32,* 458-464.

[63] Chiodo, L. M. and Covington, C. (2007). Blood lead levels and specific attention effects in young children. *Neurotoxicol. Teratol., 29,* 538-546.

[64] Nigg, J. T. and Knottnerus, G. M. (2008). Low blood lead levels associated with clinically diagnosed attention-deficit/ hyperactivity disorder and mediated by weak cognitive control. *Biol. Psychiatry., 63,* 325-331.

[65] Liu, J. X. and Han, D. (2010). Lead Affects Apoptosis and Related Gene XIAP and Smac Expression in the Hippocampus of Developing Rats. *Neurochem. Res., 35,* 473-479.

[66] Li, Y. and Xu, X. J. (2008). Determination of meconium lead level of newborn by graphite furnace atomic absorption spectrometry. *Spectroscopy and Spectral Analysis, 28,* 447-449.

[67] Li, Y. and Xu, X. (2008). Monitoring of Lead Load and its Effect on Neonatal Behavioral Neurological Assessment Scores in Guiyu, an Electronic Waste Recycle Town in China. *Journal of environment monitoring, 10,* 1233-1238.

[68] Chen, A. M. and Dietrich, K. N. (2011). Developmental Neurotoxicants in E-Waste: An Emerging Health Concern. *Environ. Health. Perspect., 119,* 431–438.

[69] Guo, Y. Y. and Huo X. (2010). Monitoring of lead, cadmium, chromium and nickel in placenta from an e-waste recycling town in China. *Science of the Total Environment, 408,* 3113-3117.

[70] Li, Y. and Huo, X. (2011). Assessment of cadmium exposure for neonates in Guiyu, an electronic waste pollution site of China. *Environ. Monit Assess, 177,* 343-351.

[71] Li, Y. and Xu, X. (2008). The hazard of chromium exposure to neonates in Guiyu of China. *Sci. Total. Environ., 403,* 99-104.

[72] Darnerud, P. O. and Eriksen, G. S. (2001). Polybrominated diphenyl ethers: occurrence, dietary exposure, and toxicology. *Environ. Health Perspect.,* 109, 49-68.

[73] Julie, B. and Herbstman, A. S. (2010). Prenatal Exposure to PBDEs and Neurodevelopment. *Environ. Health Perspect., 118,* 712-719.

[74] Wu, K. S. and Xu. X. J. (2010). Polybrominated Diphenyl Ethers in Umbilical Cord Blood and Relevant Factors in Neonates from Guiyu, China. *Environ. Sci. Technol., 44,* 813-819.

[75] Wu, K. and Xu, X. (2011). In utero exposure to polychlorinated biphenyls and reduced neonatal physiological development from Guiyu, China. *Ecotoxicology and Environmental Safety,* DOI:10.1016/j.ecoenv.2011.07.038.

[76] Chen, S. J. and Ma, Y. J. (2009). Brominated flame retardants in children's toys: concentration, composition, and children's exposure and risk assessment. *Environ. Sci. Technol.,* 43, 4200-4206.

In: E-Waste: Management, Types and Challenges
Editors: Yuan Chun Li and Banci Lian Wang
ISBN 978-1-61942-217-9

Chapter 6

RECOVERY OF VALUABLE METALS FROM FLAT PANEL DISPLAYS OF SPENT LIQUID CRYSTAL AND PLASMA TELEVISIONS

***Katsutoshi Inoue*[1], *Biplob Kumar Biswas*[1,2], *Hidetaka Kawakita*[1] and *Keisuke Ohto*[1]**
[1]Department of Applied Chemistry, Saga University, Japan
[2]Department of Applied Chemistry and Chemical Engineering, Jessore Science and Technology University, Bangladesh

ABSTRACT

Recovery processes of indium from waste liquid crystal display panels and that of silver from display panels of spent plasma televisions were developed on the basis of hydrometallurgical technology. In the former case, the powdered sample was leached with 3 M HCl and extracted metals like indium, aluminum, tin and so on were mutually separated and recovered by passing through two columns: the first column was packed with porous resins impregnated with Aliquat 336, a commercially available solvent extraction reagent of quaternary ammonium compound, while the second column was packed with those impregnated with Cyanex 923, also commercially available solvent extraction reagent of trialkylphosphine oxide. In the first column, tin, iron and zinc were removed from the leach liquor while, in the second column, only indium was selectively recovered. The metal ions trapped in these columns were eluted with 0.1 M H_2SO_4 concentrated more than about 10 times.

On the other hand, in the latter case, metals including silver were leached from powdered scrap of plasma TV monitors using acidothiourea solution, which achieved almost complete extraction of silver. The recovery of silver from acidothiourea leach liquor was achieved by means of simple cementation method by using zinc powder. Moreover, precipitation with sodium sulfide was found to be a promising means for silver recovery. Persimmon tannin solution, a natural and nonconventional coagulant, containing polyphenolic groups, was also found to completely recover silver from silver-containing acidothiourea leach liquor through coagulation-precipitaion followed by reduction of silver.

INTRODUCTION

Due to the rapid spread of liquid crystal TV in recent years, the demands for indium is now increasing as the feed material for indium tin oxide (ITO) transparent electrode, produced by means of sputtering. Since indium is contained in sphalerite in small amount, it is produced as a byproduct in the smelting of zinc. The resources of indium are limited on the earth and the countries or areas which can provide indium are also limited, which resulted in the rapid increase in the price of indium in recent years.

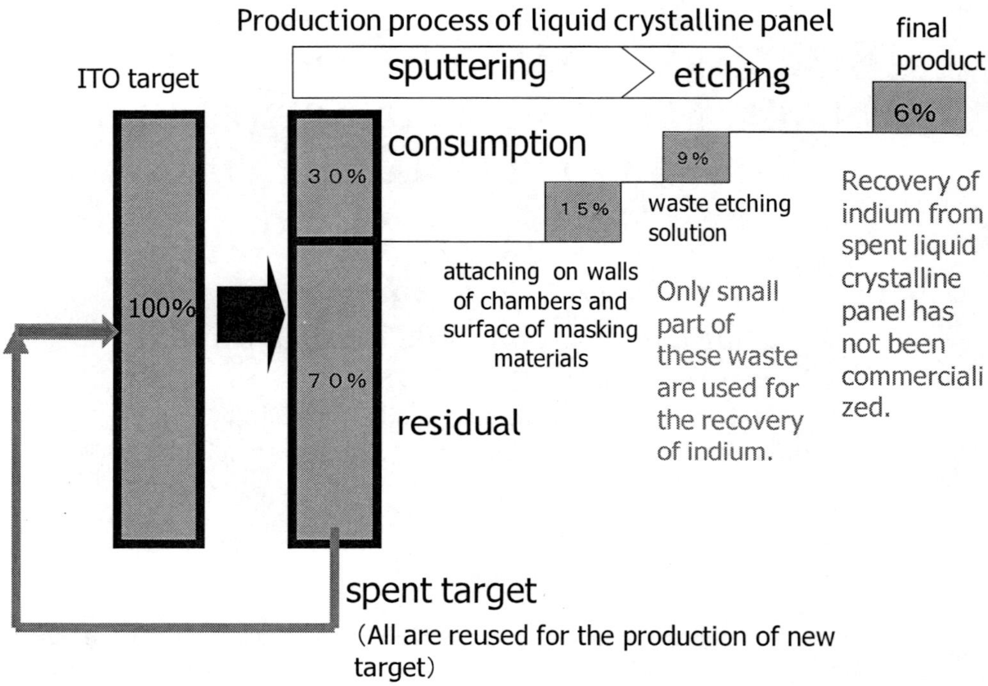

Figure 1. Mass balance of indium in the production of liquid crystalline panel [1].

Although big efforts have been devoted to the development of new technology for substituting indium with common metals such as zinc, it is expected to take a long time for its commercialization. Consequently, only one choice to avoid the shortage of indium is its effective recovery from various indium-containing wastes and its recycle.

Figure 1 shows the mass balance of indium in the production of liquid crystalline panel in Japan. [1] Although spent target generated in the production of ITO transparent electrode by sputtering is now fully recycled, indium contained in flat panel display (FPD) of liquid crystal TV has not been attempted to be recovered for recycling yet. In our previous work [2, 3], we conducted a fundamental investigation on the hydrometallurgical recovery process of indium from waste FPD.

Only small portion of indium is recovered from ITO waste attached on walls of chambers and surface of masking materials together with that in waste etching solution at present while the recovery from spent liquid crystalline panels has not been practiced yet. Figure 2 shows the flowsheet of the recovery process of indium from ITO waste attached on walls of chambers and surface of masking materials together with that in waste etching solution, which is nearly the same process for the refining of indium from ores. We developed a novel hydrometallurgical recovery process of indium from spent liquid crystalline panels as shown in Figure 3.

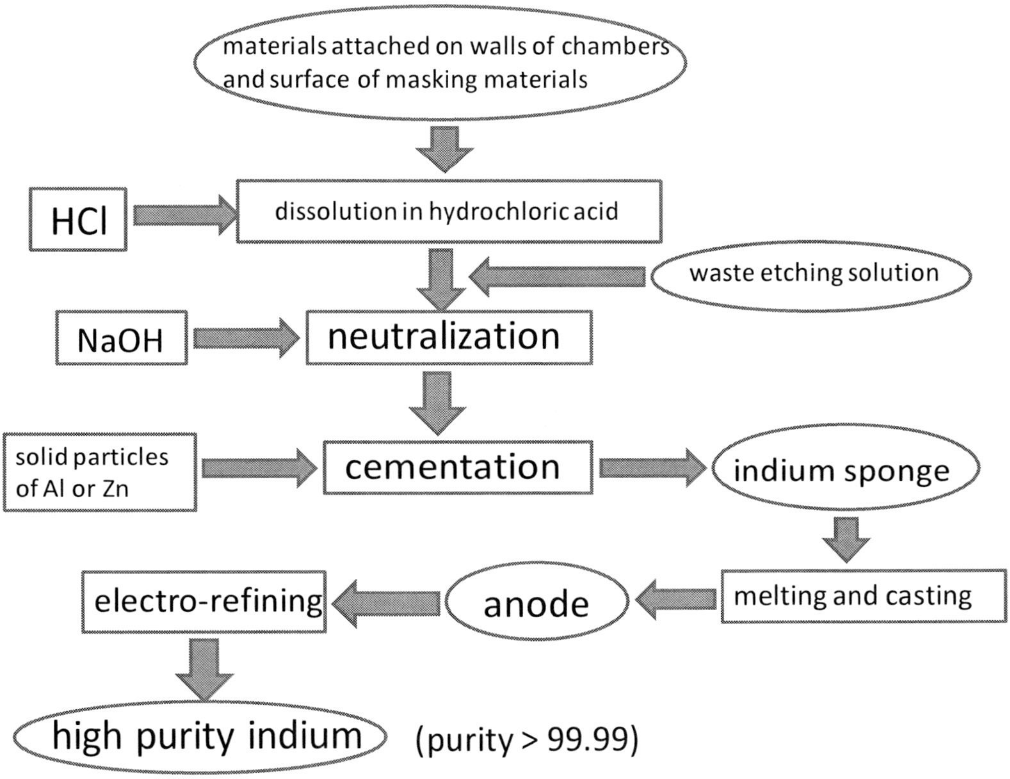

Figure 2. Flow sheet of the conventional recovery process of indium from ITO waste attached on walls of chambers and surface of masking materials together with that in waste etching solution [1].

On the other hand, silver, one of precious metals, is one of the extensively used elements in industry for various processing and in jewelry for ornament. The annual demand for silver in Japan alone is reported to be 2597 ton, of which 567 t is for electrical and electronic products [4]. The wide range of application and growing demand for this metal are the two main reasons for its recovery. Although majority of silver had been recovered from scraps such as X-ray films, photographic films and jewelry in former days, [5] spent electronic equipments such as spent personal computers have been examined in recent years to estimate the amount of potentially recoverable metals, which includes precious metals (silver, gold, palladium) and also other metals. [4]

Because some amount of silver is used as silver electrode on panels of plasma TV, scrap of plasma TV monitors is one of the potential sources of silver and there is no reason which impedes the recovery of this element. The conventional recovery process of silver as well as gold is cyanide process. According to this process, it is leached from feed materials like ores with alkaline cyanide solution and recovered by means of adsorption on activated carbon, which is further incinerated to finally recover silver as ash. Although this process has been employed for a long time extensively all over the world, it is needless to mention that the process using very toxic cyanide is not environmentally benign and should be alternated by more environmentally benign processes. Therefore, there is an urgent need to develop efficient technologies to recover this precious metal in an environmentally benign way. Consequently, we attempted to investigate an environmentally benign hydrometallurgical recovery process of silver from panels of spent plasma TV monitors in our previous work. [6]

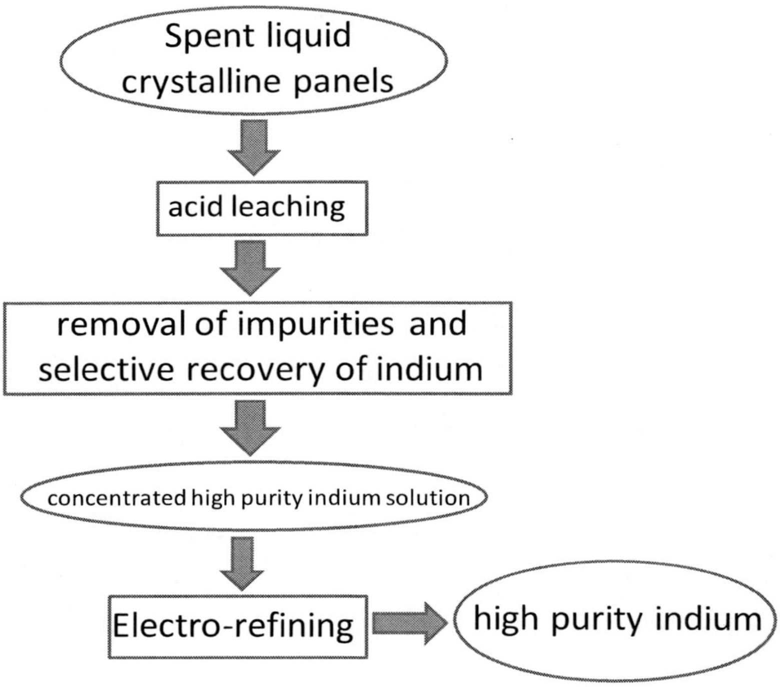

Figure 3. Flow sheet of the recovery process of indium from spent liquid crystalline panels proposed in the present work.

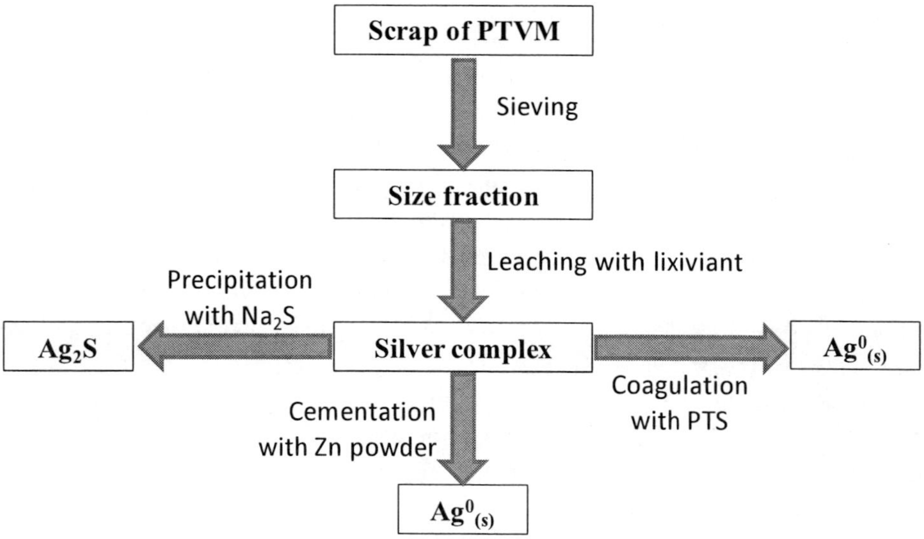

Figure 4. Flow sheet of the recovery of silver from scrap of plasma TV monitors.

In the present article, we also introduce the recovery processes from panels of spent plasma TV monitors we developed. Figure 4 shows a simplified flow sheet of the recovery process where leaching with acidothiourea is followed by any of the three methods: precipitation with Na_2S or cementation with Zn powder or coagulation-precipitaion with persimmon tannin solution. Such novel hydrometallurgical process combined with any of those three methods would certainly give solution not only from the point of view of silver recovery but also from the view point of electronic-waste management.

RECOVERY OF INDIUM FROM FPD OF SPENT LIQUID CRYSTALLINE TV

Leaching of Metal Values from FPD of Spent Liquid Crystalline TV

The actual waste sample of FPD of liquid crystal TV produced by Samsung Electronics Co. Ltd., Korea, was supplied from Shimada Shokai Co. Ltd., Japan. It was crushed by hammer into small particles. One g of particles was mixed with 10 mL of aqua regia, 3 M (M = mol/L) hydrochloric acid, sulfuric acid and nitric acid to be shaken for a required time in order to screen the most suitable lixiviant for this sample. After shaking, sample solutions were taken for the analysis of metal concentration by using Shimadzu model ICPS-8100 ICP/AES spectrometer.

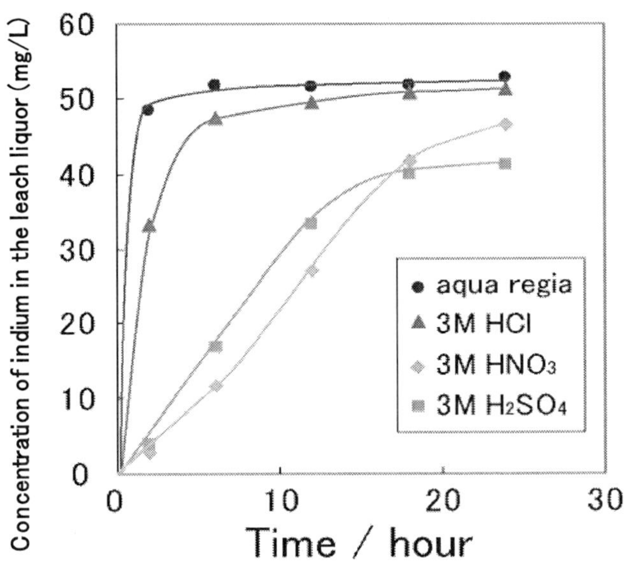

Figure 5. Leaching of indium from powder of crushed spent FPD with various acids. Solid/liquid ratio = 1 g/10 mL.

Figure 5 shows the relationship between the concentration of indium(III) in the various leach liquors and the leaching time. From this figure, indium(III) is the most effectively leached with aqua regia; however, nearly the same leaching behavior was observed with 3 M hydrochloric acid while effective leaching was not observed with 3 M nitric acid and sulfuric acid. On the other hand, total dissolution of the powdered sample in aqua regia suggested the metal contents (mg/L) as follows: Al 230, In 205, Ba 175, Si 105, Ca 95, Fe 60, Zn 39.5, Cr 29, Sn 24, Mg 10.

Screening of the Solvent Extraction Reagent of Indium(III) from Chloride Media

In order to selectively recover indium(III) from the above-mentioned chloride leach liquor, solvent extraction behaviors of indium(III) from varying chloride concentration by using 5 kinds of reagents, TOPO (trioctylphosphine oxide), Cyanex 923, TBP (tribytylphosphate), Aliquat 336, and TIOA (triisooctylamine), the chemical structures of which are shown in Figure 6 were observed to screen the most suitable reagent for this purpose.

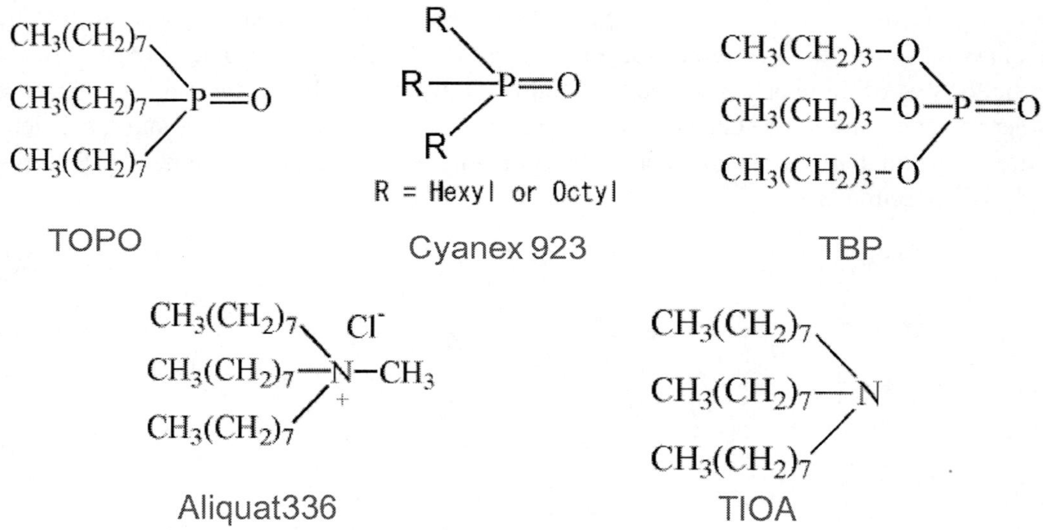

Figure 6. Chemical structures of the reagents for the solvent extraction of indium(III) employed in the present work.

Figure 7 shows the plot of % extraction of indium(III) from varying concentrations of hydrochloric acid with some solvent extraction reagents in toluene. From this figure, it is seen that Cyanex 923, the mixture of 4 kinds of trialkylphosphine oxides with different alkyl radicals, exhibits the highest extraction though TOPO also exhibits the similar extraction behavior. On the other hand, TBP exhibited no extraction of indium(III) over the whole concentration region of hydrochloric acid.

Figure 8 shows the % extraction of indium(III) with Cyanex 923 in toluene from varying concentrations of hydrochloric acid and sulfuric acid as well as that containing 0.2 M ammonium chloride. It is seen that although indium(III) is not extracted from sulfuric acid the addition of some amount of chloride improves the extraction similar to hydrochloric acid, suggesting that chloride ion plays an important role in this extraction system and that sulfuric acid can be employed as stripping agent for indium(III).

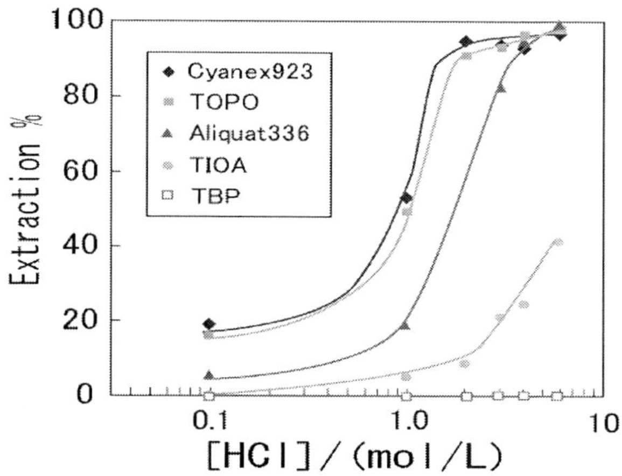

Figure 7. Plot of % extraction of indium(III) with some reagents in toluene against concentration of hydrochloric acid. [In(III)] = 4 mM, [extractant] = 80 mM. Temperature = 303 K.

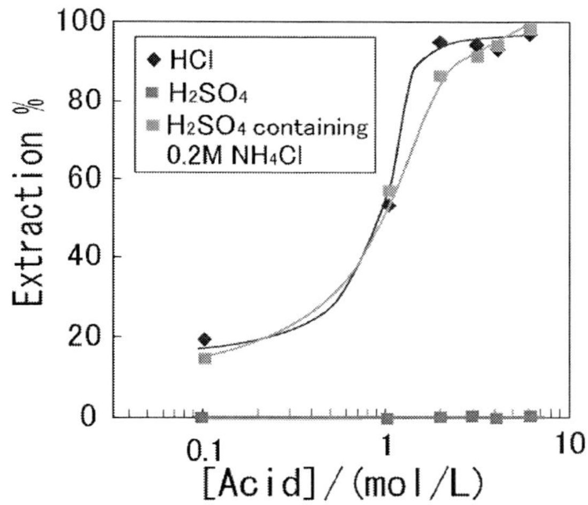

Figure 8. % extraction of indium(III) with Cyanex 923 in toluene from hydrochloric acid and sulfuric acid. [In(III)] = 4 mM, [extractant] = 80 mM. Temperature = 303 K.

Adsorptive Recovery of Dilute Concentration of Indium(III) Using Solvent Impregnated Porous Resins

As mentioned above, solvent extraction reagents of trialkylphosphine oxide like Cyanex 923 and TOPO are suitable to recover indium(III) from acidic chloride media, the concentration of indium(III) in the present case (205 mg/L) is too low to employ solvent extraction technique. Consequently, we employed adsorption technique by using a porous resin impregnated with Cyanex 923, which was prepared as follows. The sample of Amberlite XAD-7, the porous spherical beads of acrylic ester resin with no functional groups, was put in acetone solution containing Cyanex 923 and shaken for 12 h, after which the beads were taken by filtration and vacuum dried to remove acetone. The contents of the impregnated Cyanex 923 in the beads were 1.29 mol/kg.

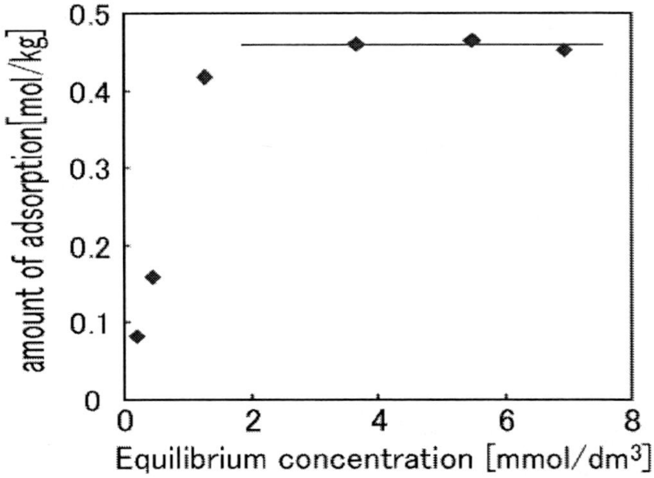

Figure 9. Adsorption isotherm of indium(III) on the XRD-7 resin impregnated with Cyanex 923 from 3 M hydrochloric acid at 303 K.

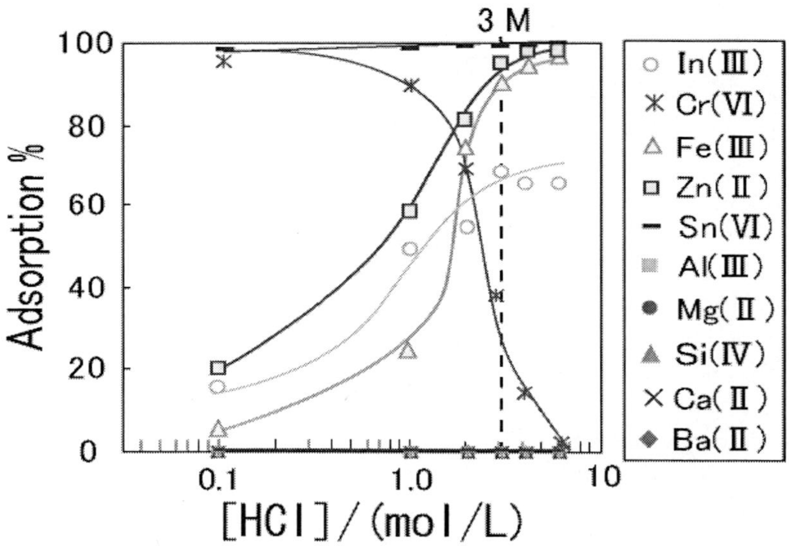

Figure 10. % adsorption of all metal ions existing in the leach liquor of the spent FPD from varying concentration of hydrochloric acid solution on the Cyanex 923 impregnated resin. Concentration of each metal ion = 0.2 mM, solid/liquid ratio = 30 mg/15 mL. Temperature = 303 K.

Figure 9 shows the adsorption isotherm of indium(III) from 3 M hydrochloric acid solution on thus prepared impregnated resin at 303 K. The amount of adsorbed indium(III) increases with increasing indium(III) concentration in the low concentration range while it tends to approach a constant value in the high concentration range. From this constant value, the maximum adsorption capacity of indium(III) on this resin was evaluated as 0.45 mol/kg.

Figure 10 shows the adsorption of metal species existing in the leach liquor on the resin impregnated with Cyanex 923 at varying hydrochloric acid concentration. From this figure, it is seen that tin(IV) is quantitatively adsorbed over the whole concentration region of hydrochloric acid while adsorption of zinc(II), iron(III) and indium(III) increases with increasing hydrochloric acid concentration. On the other hand, that of chromium(VI) decreases with increasing hydrochloric acid concentration and those of calcium(II), magnesium(II), barium(II), aluminum(III) and silicon(IV) are negligible. Although indium(III) is effectively adsorbed on this resin, tin(IV), iron(III) and zinc(II) are more effectively or selectively adsorbed over indium(III) from 3 M hydrochloric acid solution in particular, which suggests that these 3 metal ions should be selectively removed from the leach liquor of 3 M hydrochloric acid solution in advance prior to the recovery of indium(III) to finally obtain high pure indium(III) solution.

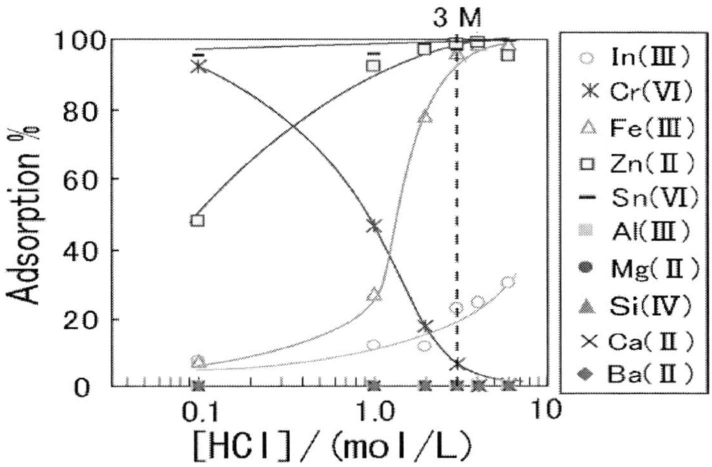

Figure 11. % adsorption of all metal ions existing in the leach liquor of the spent FPD from varying concentration of hydrochloric acid solution on the Aliquat 336 impregnated resin. Concentration of each metal ion = 0.2 mM, solid/liquid ratio = 30 mg/15 mL. Temperature = 303 K.

For this purpose, these 3 metal ions were tried to be selectively removed by using another types of resin, Amberlite XAD-7 resin impregnated with Aliquat 336, trioctylmethyl ammonium chloride, a quaternary amine type of solvent extraction reagent, which was prepared in the similar manner with the resin impregnated with Cyanex 923.

Figure 11 shows the adsorption behaviors of metal species in question on the resin impregnated with Aliquat 336 at varying hydrochloric acid concentration. From this figure, it is seen that although % adsorption of these 3 impurity metal ions is greater than 90 %, that of indium(III) is less than 20 % at 3 M hydrochloric acid in the case of Aliquat 336 impregnated resin, suggesting that this resin is suitable for selectively removing these impurity metal ions from indium(III). On the other hand, adsorption of other metal ions, aluminum(III), magnesium(II), silicon(IV), calcium(II), and barium(II), is negligible also on this resin.

Adsorptive Recovery and Purification of Indium by Means of Continuous Operation Using Columns Packed with Two Types of Impregnated Resins

On the basis of the results of the batchwise adsorption tests mentioned above, continuous adsorption test using two columns packed with 0.60 g of each of two types of impregnated resins connected in series as shown in Figure 12 was carried out to obtain highly purified and concentrated indium(III) solution. In this continuous adsorption test, synthetic test solution of 3 M hydrochloric acid containing the same concentration of each metal ion measured in the leaching test was fed to the columns at the feed rate of 5.0 mL/h with a peristaltic pump.

Figure 13 shows the breakthrough profiles of all metal ions tested from the first column shown in Figure 12 packed with the resin impregnated with Aliquat 336. Although iron(III), zinc(II) and tin(IV) were not detected in the effluent from the column even after 100 bed volumes (B.V.), aluminum(III), magnesium(II), silicon(IV), calcium(II), barium(II), and chromium(VI) were immediately broken through just after the initiation of the feed. On the other hand, indium(III) was gradually broken through until 50 B.V., suggesting that some amount of indium(III) is retained in this column.

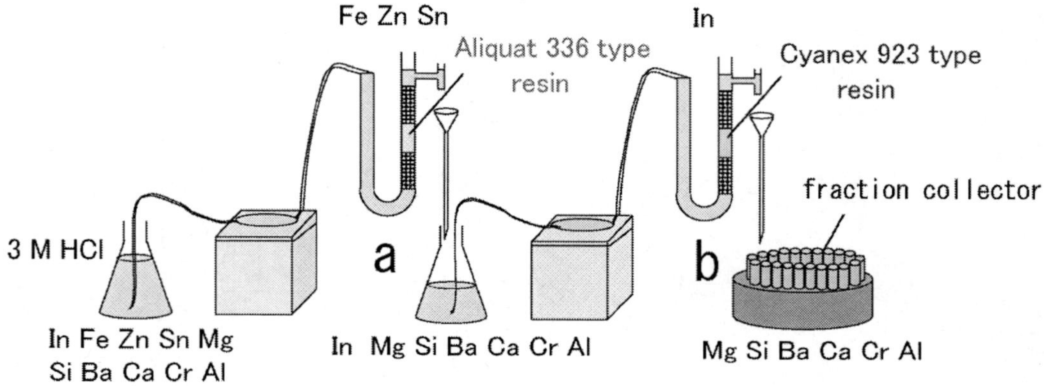

Figure 12. Experimental set up for the continuous adsorption test for the selective recovery of indium(III).

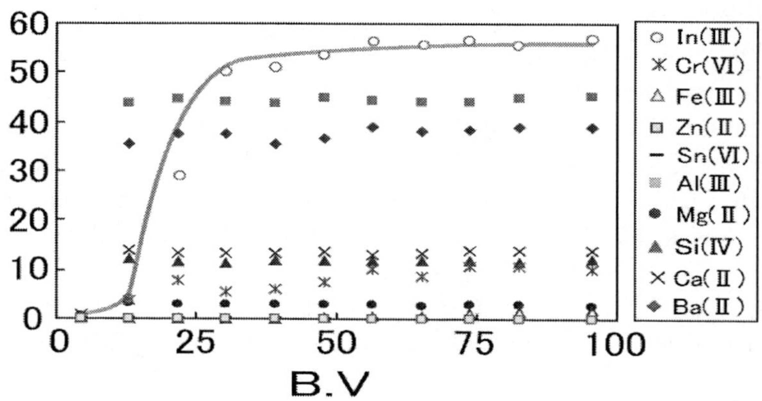

Figure 13. Breakthrough profiles (relationship between effluent metal concentrations (mg/L) and bed volume corresponding to the time from the start of the feed) of all metal ions tested from the first column packed with Aliquat 336 impregnated resin.

Figure 14 shows the breakthrough profiles from the second column packed with Cyanex 923 impregnated resin together with the elution profile of indium(III) (after 50 B.V.) from the same column. All of aluminum(III), magnesium(II), silicon(IV), calcium(II), barium(II), and chromium(VI) which had passed through the first column without being adsorbed were immediately broken through just after the initiation of the feed also in the second column. On the other hand, indium(III) was not detected at all in the effluent from the second column up to 50 B.V., suggesting that indium(III) is successfully purified free from the contamination of other metal ions. As mentioned earlier, after the 50 B.V., the second column was washed by water and, subsequently, with 0.1 M sulfuric acid. As seen from the elution profile of indium(III) in Figure 14, all indium(III) adsorbed in the second column is eluted enriched more than 10 times compared to the feed solution.

From the above-mentioned result, it can be concluded that small amounts of indium can be effectively extracted with 3 M hydrochloric acid from crushed spent FPD and can be purified and enriched by passing through the columns packed with porous resins impregnated with the solvent extraction reagents Aliquat 336 and Cyanex 923.

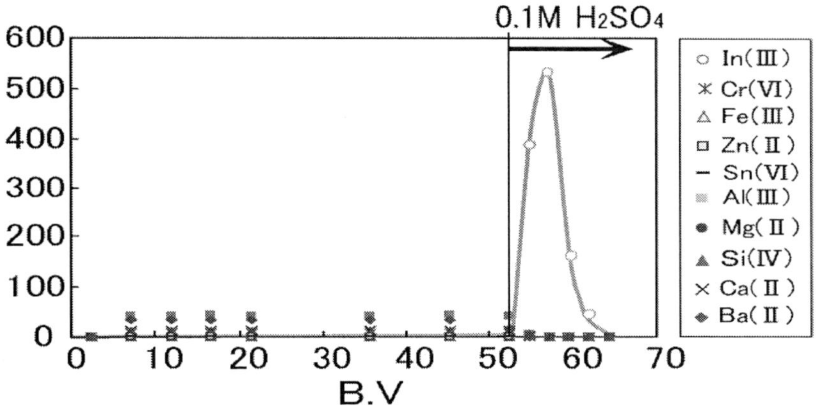

Figure 14. Breakthrough profiles (relationship between effluent metal concentrations (mg/L) and bed volume corresponding to the time from the start of the feed) (up to 50 B.V.) from the second column packed with Cyanex 923 impregnated resin and the elution profile of indium(III) (after 50 B.V.) from the same column.

RECOVERY OF SILVER FROM PANELS OF SPENT PLASMA TV

Characteristics of the Scrap of Spent Plasma TV Monitors and Determination of Silver Content

The sample of crushed scrap of spent plasma TV monitors was kindly provided by Nishinihon Kaden Recycle Corporation (NKRC), Kitakyushu, Japan. The average moisture content present in the as-received scrap of spent plasma TV monitors sample was found to be 7 wt.% while the specific gravity was determined to be 1.88. For the determination of silver content in the sample, total dissolution experiments were carried out by using aqua regia and HCl at their boiling temperatures as follows. One gram of the sample was taken together with 30 ml of digesting reagent (either aqua regia or concentrated HCl) in a 100 ml beaker and stirred at boiling temperature for about 30 min. The insoluble residue was double baked to ensure a complete dissolution. The digestion liquid was diluted with deionized water and filtered. Finally, the contents of all inorganic elements in the filtrate were measured by using ICP/AES spectrometry as listed in Table 1.

Table 1. An analysis of metals (mg/g) present in the scrap of spent plasma TV monitors

Elements	Dissoluting reagents	
	HCl	Aqua regia
Na	2.60	2.96
K	4.08	6.12
Ca	8.16	6.63
Fe	1.79	1.43
Ni	0.12	0.16
Cu	0.20	0.37
Zn	7.14	7.14
Ag	2.29	2.35
Al	0.37	0.27
Si	5.61	3.21
Pb	4.49	4.23
Au	0.15	0.18

Leaching of Silver with Acidic Thiourea Solution

Since thiourea is known to form a stable cationic complex with silver in the form of $Ag[CS(NH_2)_2]_3^+$, it represents a very attractive leaching alternative. [7,8] Very high extent of silver extraction was achieved by using thiourea together with dilute sulfuric acid solution as shown in Figure 15. That is, silver in the scrap was found to be successfully leached out by using a mixture of 0.05 M thiourea and 0.05 M H_2SO_4, which provides an important information about efficient silver extraction by using dilute acidothiourea solution, which indicated a clear advantage over other methods, such as that where the addition of ferric ions is required. [9] It is obvious from Figure 15 that effective leaching is achieved with a combination of low concentration of acid and low concentration of thiourea. Even though, dilute sulfuric acid alone has only very little effect on silver extraction, the mixture of sulfuric acid and thiourea offers excellent leaching performance. This is because low concentration of sulfuric acid facilitates complexation between silver and thiourea while high concentration acid impedes the complexation by protonating thiourea, which results in low silver extraction.

Silver Recovery by Means of Cementation Using Zinc Powder

A cementation test of silver was carried out for silver-containing acidothiourea leach liquor, where a measured weight of zinc powder was added to 20 ml of the leach liquor and shaken at 150 rpm at 30°C. The solution was then decanted and filtered. The filtrate was analyzed to measure the residual silver concentration. Silver cementation is an inexpensive and simple way to deposit thin metal films [10]. Difference in oxidation-reduction potential (ORP) of silver (+ 0.80 volt) and zinc (– 0.76 volt) may lead to cementation of silver from silver-containing acidothiourea leach liquor. In order to elucidate the cementation reaction of silver complex with zinc powder, various dosages of zinc powder (< 75 μm) were tested, as shown in Figure 16. As seen from this figure, only a small dose of zinc powder was found to be effective for the recovery of silver from silver-containing acidothiourea leach liquor and an optimum solid/liquid ratio was determined to be 5 mg/ml. In the present case, silver can be cemented by zinc powder according to the following reaction.

$$2Ag[CS(NH_2)_2]_3^+ + Zn^0_{(s)} \rightarrow 2Ag^0_{(s)} + Zn \cdot thiourea \cdot complex \tag{1}$$

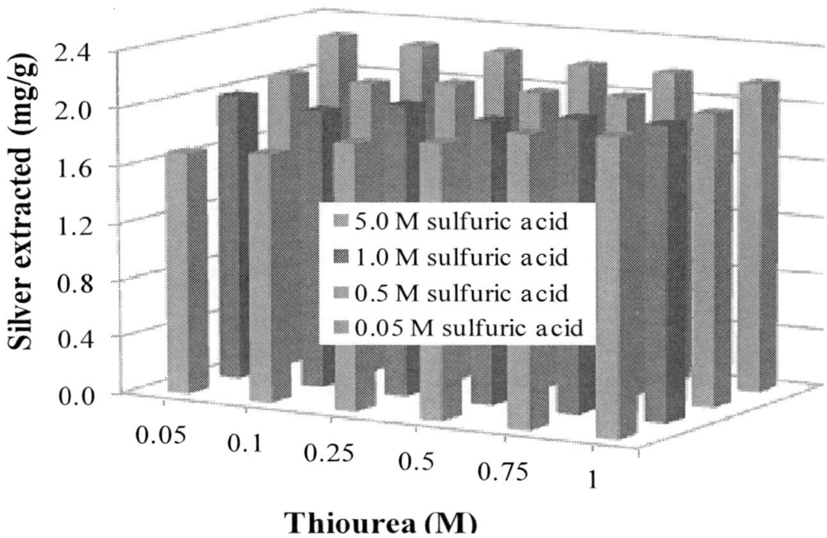

Figure 15. Leaching of silver from scrap of spent plasma TV monitors by using varying concentrations of acidic thiourea. Conditions: liquid/solid ratio = 250 ml/g, T = 30°C, shaking time = 4 h.

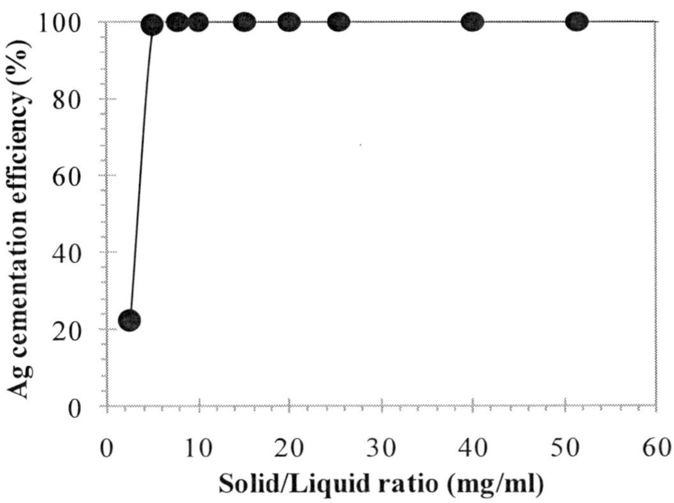

Figure 16. Recovery of silver by cementation from silver-containing acidothiourea leach liquor at varying dosages of zinc powder. Conditions: silver-containing acidothiourea leach liquor = 20 ml, T = 30°C.

PRECIPITATION OF SILVER WITH SODIUM SULFIDE

Another approach of silver recovery was tested by a precipitation technique with sodium sulfide solution. It provides a practical route to precipitative recovery of silver as its sulfide. Varying concentrations of sodium sulfide ($Na_2S{\cdot}9H_2O$) were added to 10 ml of silver-containing acidothiourea leach liquor and shaken for a couple of minutes, and then allowed to settle for 24 h. The supernatant was then collected and the concentration of silver was measured. As shown in Figure 17, very small dose (0.05 ml/ml) of both 20 mM and 50 mM of Na_2S accomplished a complete (100%) silver recovery through precipitation. Such a high degree of precipitation suggests a possible recovery route for silver from silver-containing acidothiourea leach liquor. Precipitated silver sulfide, however, needs further treatment to produce metallic silver by reduction, which will take a high cost.

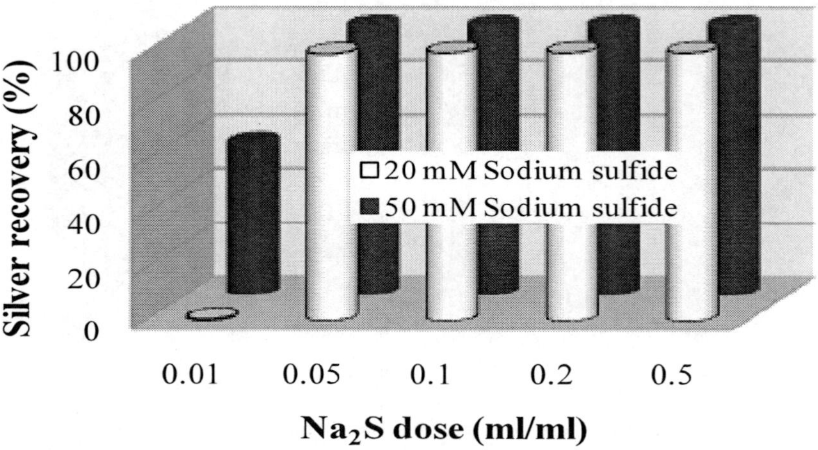

Figure 17. Recovery of silver through precipitation using sodium sulfide. Conditions: Silver-containing acidothiourea leach solution = 10 ml, Shaking time = 2 min, Settling time = 24 h, T = 30°C.

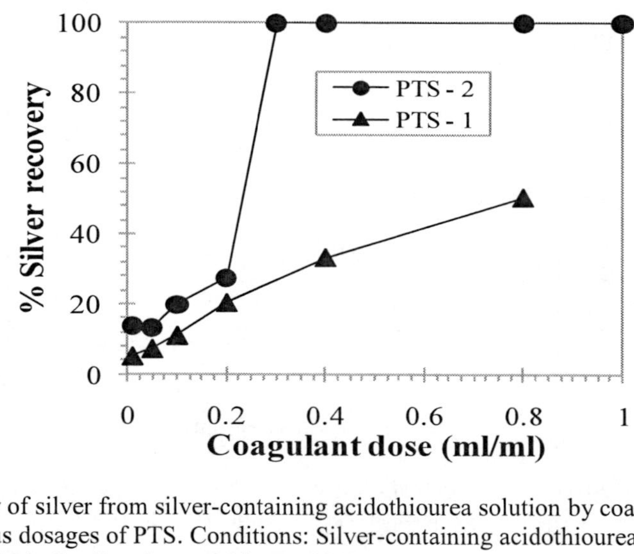

Figure 18. Recovery of silver from silver-containing acidothiourea solution by coagulation-precipitation method using various dosages of PTS. Conditions: Silver-containing acidothiourea leach solution = 10 ml, Shaking time = 12 h, Settling time = 24 h, T = 30°C.

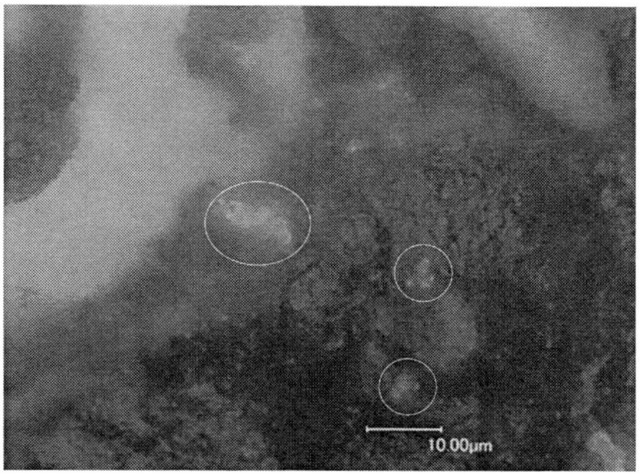

Figure 19. Digital microscopic image (4000 times magnification) of precipitation cake and observation of metallic silver.

Coagulation-Precipitation of Silver from Silver-Containing Leach Liquor by Using Persimmon Tannin Solution

Coagulation-precipitation is well-known as a useful tool for the recovery of metals from a solution after leaching. In the present work, a nonconventional coagulant, astringent persimmon tannin extract, was investigated for silver recovery. Persimmon tannin contained in persimmon extract has a large number of functional groups of catechol and pyrogallol, which function as strong reducing agents. Persimmon tannin solution (PTS) was prepared by dissolving a measured amount of persimmon tannin extract (containing 19.5% tannin) in deionized water (at pH 7) and in alkaline solution adjusted to pH 13.8 by adding a small amount of sodium hydroxide, which were termed as PTS-1 and PTS-2, respectively, hereafter. Measured amount these coagulants (PTS-1 and PTS-2) were separately mixed with silver-containing leach liquor for the silver recovery. The mixture was incubated for a required time and then filtered. The formation of silver particles during the coagulation was confirmed by digital microscopy (KEYENCE VHX-1000). The silver concentrations in the filtrate were determined by ICP/AES spectrometry.

Figure 18 shows the % recovery of silver at various dosages of PTS. Silver recovery increases with the increase in coagulant dose in both cases of PTS-1 and PTS-2. However, it is evident that PTS-2 is very efficient for the recovery of silver since 100% recovery was achieved at a coagulant dose higher than 0.3 ml/ml. PTS-1, on the other hand, gave the recovery only 50% even at a high coagulant dose, for example, even at 0.8 ml/ml. This is attributed to the fact that powdery persimmon tannin extract, which dissolved completely at pH 13.8 (PTS-2), allowed all polyphenolic groups to be dissolved in the solution, thus enabling coagulation with silver-thiourea complex. In contrast, persimmon tannin extract was only partially dissolved at pH 7 (PTS-1) and did not enhance silver recovery through coagulation-precipitaion beyond 50%. It should be noted that persimmon tannin contains a large number of functional groups of catechol and pyrogallol, which function as strong reductive agents for metal ions. [11-13]

In this study, a reduction of silver was postulated since metallic silver was found on the surface of the precipitation cake as shown in Figure 19. This phenomenon is attributed to a strong reducing capability of the above-mentioned polyphenolic functional groups as expressed by Eqations 2 and 3.

$$R-OH \rightarrow R=O+H^{+}+e^{-} \tag{2}$$

$$Ag^{+}+e^{-} \rightarrow Ag^{0}_{(s)} \tag{3}$$

CONCLUSION

From the recovery test of indium from scrap of spent FPD, followings can be concluded.

1) Indium can be leached with 3 M HCl.
2) Impurities in the leach liquor which should be removed are aluminum, silicon, calcium, iron, zinc, magnesium, tin, barium and chromium.
3) Recovery of indium and removal of impurities can be achieved by using adsorption columns packed with solvent impregnated resins which exhibit high selectivity to indium and impurities, respectively.
4) The major impurities which impedes the recovery and purification of indium are iron, tin and zinc.
5) These impurities can be removed by the resin impregnated with Aliquat 336 prior to the recovery of indium.
6) Indium can be recovered in high concentration by the resin impregnated with trialkylphosphine oxide.
7) The elution of indium can be achieved by dilute sulfuric acid.

On the other hand, from the recovery test of silver from scrap of spent plasma TV monitors, followings can be concluded.

8) Almost complete extraction of silver in scrap of spent plasma TV monitors was achieved by using dilute acidothiourea leach liquor (mixture of 0.05 M thiourea and 0.05 M sulfuric acid) at 30°C and at liquid to solid ratio higher than 100 ml/g in 4 h.
9) Persimmon tannin extract dissolved in alkaline solution at pH 13.8 was very efficient for silver recovery from acidothiourea leach solutions. 100 % silver recovery was achieved at coagulant dose higher than 0.3 ml/ml and metallic silver was found to be formed at the surface of the precipitation cake.
10) Precipitation with sodium sulfide was also found to be very effective since both 20 and 50 mM of sodium sulfide dose of 0.05 ml/ml yielded 100% recovery of silver.
11) The cementation study by using zinc powder indicated that 100% of silver was cemented at an optimum solid/liquid ratio of 5 mg/ml.

12) The investigated process based on acidothiourea leaching avoided environmental problems because of its nontoxic nature, and indicated an ample opportunity of silver recovery through coagulation-precipitation and cementation techniques.

REFERENCES

[1] H. Senzaki, *Preprint of Annual Meeting of MMIJ* (Mining and Materials Processing Institute of Japan), Paper No. F-6 (2007) (in Japanese)

[2] K. Inoue, M. Nishiura, H. Kawakita, K. Ohto, H. Harada, *"Recovery of indium from spent panel of liquid crystal display panel",* Kagakukogaku Ronbunshu, 34, 282-286 (2008) (in Japanese)

[3] K. Inoue, M. Nishiura, H. Kawakita, K. Ohto, H. Harada, "Recovery of indium from spent flat-panel display with solvent impregnated resins", In: *"Solvent Extraction: Fundamentals to Industrial Applications* (Proceedings of ISEC 2008 International Solvent Extraction Conference, Sep. 15-19, 2008, Tucson, AZ, USA), vol.1, pp.317-322 (2008)

[4] T. Shirahase, A. Kida, "Metals content in one waste personal computer by detailed dismantling*", Material Cycles & Waste Management Research*, 20, 217-230 (2009) (in Japanese)

[5] V. I. E. Ajiwe, I.E. Anyadiegwu, "Recovery of silver from industrial wastes, cassava solution effects", *Separation & Purification Technology*, 18, 89-92 (2000)

[6] K. Inoue, B. K. Biswas, H. Kawakita, K. Ohto, A. Hoshino, *"Recovery of silver from spent plasma TV monitors",* In: Proceedings of TMS 2011, 140th Annual Meeting & Exhibition (Late News Proceedings), Feb.27-Mar.3, 2011, San Diego, CA, USA), pp.229-237 (2011)

[7] P. Baláz, J. Ficeriová, V. Šepelák, R. Kammel, "Thiourea leaching of silver from mechanically activated tetrahedrite", *Hydrometallurgy*, 43, 367 – 377 (1996).

[8] J. Li, J. Miller, "A review of gold leaching in acid thiourea solutions", *Mineral Processing and Extractive Metallurgy Review*, 27, 177 – 214 (2006).

[9] C.K. Chen, T.N. Lung, C.C. Wan, "A study of the leaching of gold and silver by acidothioureation", *Hydrometallurgy*, 5, 207 – 212 (1980).

[10] M. Fbián, P. Baláž, J. Briančin, "Study of the silver ions cementation after mechanical activation of cementator", *Hydrometallurgy*, 97, 15-20 (2009).

[11] H. Kawakita, R. Yamauchi, D. Parajuli, K. Ohto, H. Harada, K. Inoue, "Recovery of gold from hydrochloric acid by means of selective coagulation with persimmon extract", *Separation Science and Technology,* 43, 2375 – 2385 (2008).

[12] Y.H. Kim, Y. Nakano, "Adsorption mechanism of palladium by redox within condensed-tannin gel", *Water Research*, 39, 1324 – 1330 (2005).

[13] Nakajima, Y. Baba, "Mechanism of hexavalent chromium adsorption by persimmon tannin gel", *Water Research*, 38, 2859 – 2864 (2004).

In: E-Waste: Management, Types and Challenges
Editors: Yuan Chun Li and Banci Lian Wang
ISBN 978-1-61942-217-9

Chapter 7

MANAGING ELECTRONIC WASTE: AN ANALYSIS OF STATE E-WASTE LEGISLATION*

Linda Luther

ABSTRACT

Pursuant to the Resource Conservation and Recovery Act (RCRA), the U.S. Environmental Protection Agency (EPA) has established regulations regarding the disposal of hazardous wastes. Although there are federal requirements under RCRA for the management of hazardous waste, some states have opted to implement *more* stringent requirements — particularly with regard to the management of certain hazardous wastes generated by households and small businesses (entities that are essentially exempt from RCRA's hazardous waste management requirements).

One category of household hazardous waste that many states are choosing to regulate more strictly is electronic waste, commonly referred to as "e-waste." E-waste generally refers to obsolete, broken, or irreparable electronic equipment like televisions, computers and computer monitors, laptops, printers, cell phones, copiers, fax machines, stereos, or video gaming systems. Cathode ray tubes (CRTs) in televisions and computer monitors have presented a particular concern to states, primarily due to the potentially significant amounts of lead they contain and the large numbers in which they are generated.

State concerns specific to the landfill disposal or incineration of e-waste are largely due to its increasing volume and often bulky nature; hazardous constituents, such as lead and mercury, it may contain; its high cost of recycling; and the inability of interested stakeholders, such as electronics retailers and manufacturers, to reach consensus on how to voluntarily implement a national e-waste management system. States have responded to this concern by enacting their own e-waste management laws. Requirements of those laws range from a ban only on the landfill disposal or incineration of designated e-wastes to the implementation of a full e-waste collection, transportation, and recycling system.

To date, 13 states have enacted some form of e-waste management law (as many as 20 states proposed e-waste laws in 2006 and 2007). Although the goal of each law is similar — to avoid landfill disposal and incineration of certain types of e-waste — approaches taken to achieve that goal differ significantly. However, most state laws and

* This is an edited, reformatted and augmented version of Congressional Research Service Report RL34147, dated September 10, 2007.

proposals have certain broad elements in common, such as specifying the electronic devices covered under the law; how a collection and recycling program will be financed; collection and recycling criteria that must be met to minimize the impact to human health and the environment; and restrictions or requirements that products must meet to be sold in the state.

As more states propose e-waste legislation, potentially regulated stakeholders (particularly electronics manufacturers and retailers) have expressed concern that they will be required to comply with a patchwork of state requirements throughout the United States. This concern has led to an increased call for federal legislation regarding e-waste management. To help policy makers better understand the impact of state e-waste legislation, this report discusses issues that have led to state action, common elements in state-waste laws and proposals, and an overview of each enacted state law.

INTRODUCTION

Pursuant to the Resource Conservation and Recovery Act (RCRA), [1] the U.S. Environmental Protection Agency (EPA) has established regulations regarding the transport, treatment, storage, and disposal of hazardous wastes. RCRA establishes certain minimum standards that states must meet. However, states have the option to implement requirements that are *more* stringent than those specified under RCRA. Many states have opted to do so — particularly with regard to the management of certain hazardous wastes generated by households.

Households and certain small businesses are essentially exempt from RCRA. [2] This means that under federal law, hazardous wastes generated by those entities may be disposed of in municipal solid waste landfills or incinerators. One category of household hazardous waste that many states are choosing to regulate more strictly is electronic waste, commonly referred to as "e-waste."

There is no universally accepted definition of e-waste, but it generally refers to obsolete, broken, or irreparable electronic equipment such as televisions, computers and computer monitors, laptops, printers, cell phones, VCRs, DVD players, copiers, fax machines, stereos, and video gaming systems. State and local agencies, particularly municipal waste management agencies, have become increasingly concerned about the landfill disposal or incineration of e-waste because of the large volumes in which it is being generated and because of the hazardous constituents the waste may contain.

To avoid landfill disposal or incineration, e-waste may be recycled. Recycling may include any of a number of services or processes, such as: sorting to find reusable devices (which may in turn be sold or donated to an entity such as a school or charitable organization); demanufacturing into component parts that can be resold; or further processing components to extract materials such as metals, glass, or plastic. [3] For example, leaded glass in a cathode ray tube (CRT) may be recovered and reprocessed to produce new CRTs. Recycling can be a costly process (see "The Cost of Recycling Electronics," below). A challenge to many states is how to finance an e-waste collection and recycling program.

To date, 13 states have enacted some form of legislation or regulations that will affect e-waste recycling and disposal practices. As more states propose such legislation, potentially regulated stakeholders (particularly electronics manufacturers and retailers) have expressed concern that they will be required to comply with a patchwork of state requirements throughout the United States.

In 2005, two congressional hearings were held to explore issues associated with e-waste, and the Congressional E-Waste Working Group was formed. One goal common to both the hearings and the establishment of the working group was to explore potential national solutions to the e-waste management issue. With increased legislative activity in the states, it is anticipated that stakeholders will increase their call for federal legislation regarding e-waste management. To illustrate the issues associated with individual state action, this report discusses the key issues that have led to state action, describes common elements in state waste laws and proposals, and provides an overview of each enacted state law.

ISSUES LEADING TO STATE ACTION

The Volume and Bulky Nature of E-Waste

The proliferation of and increasingly rapid technological advances in electronics means that the volume of e-waste generated in the United States is large and growing. Until recently, data regarding electronic products sold, stored, recycled, disposed of, and exported in the United States were limited. In 2007, EPA completed a study that attempted to gather more data. [4] According to that study, as of 2005, of electronic products sold in the United States between 1980 and 2004,

- almost half (976 million units) were still in use or reuse,
- almost 42% (842 million units) were recycled or disposed of, and
- almost 9% (180 million units) were in storage.

Further, in 2005 alone, EPA estimated that between 1.9 to 2.2 million tons of electronics became obsolete. Of that amount, between 1.5 to 1.9 million tons were discarded, primarily in landfills. Although EPA estimates that e-waste comprises about 2% of the municipal solid waste stream, it is anticipated that this percentage will grow as consumers continue to replace old and outdated electronic equipment and discard equipment in storage. This will be the case particularly after the transition from analog to digital television broadcasts [5] and with the increased use of flat-screen televisions and computer monitors.

In addition to the bulky nature of electronic devices such as televisions and computers, the increasing volume of e-waste concerns some states, particularly state and municipal waste management agencies. Because these items have the potential to be reused or recycled, some states have become increasingly interested in diverting such waste from municipal landfills.

Hazardous Components in E-Waste

The potential presence of various toxic or hazardous components is another reason that e-waste is a concern. For example, cathode ray tubes (CRTs), [6] computer central processing units (CPUs), and other electronic devices generally contain significant quantities of lead. [7]

CRTs contain an average of four pounds of lead but may contain more, depending on the size, age, and make of the device. Electronic devices are also likely to contain a number of other heavy metals, such as mercury, beryllium, barium, chromium, nickel, or zinc. [8] Also, brominated flame retardants are commonly added to the plastic housing of televisions, computers, and other electronic devices. [9] When disposed of in landfills or incinerated, hazardous components of e-wastes may be released into the groundwater or air. [10]

In some instances, hazardous materials are used in electronics to make those devices safe for consumer use or because less toxic alternatives are not readily available. For example, CRT glass is infused with lead to protect users from radiation emitted from the tube; flame retardants are added to plastics to prevent the heated electronic devices from catching fire. Although flat panel monitors are replacing CRTs, those monitors often need mercury to operate efficiently. The continued use of certain hazardous substances makes the need for recycling options greater, if the goal is to minimize the disposal of those substances in a landfill or incinerator.

The Cost of Recycling Electronics

There are various elements that contribute to the cost of a state recycling program. The recycling infrastructure itself includes the cost of collecting, transporting, and sorting the devices. There is also a cost associated with recycling the devices themselves. Recyclers and refurbishers often charge a fee for their services because their costs outweigh the revenue received from recycled commodities (e.g., glass, metals, or reusable components) or from the sale of refurbished units. Although they will recover a certain amount of usable scrap from e-waste, they will also likely incur expenses when they have to handle and dispose of any hazardous components. Also, unlike household consumers, recyclers will be regulated under RCRA and are subject to the more stringent requirements applicable to hazardous waste storage, transportation, and disposal.

Some stakeholders argue that one method of lowering the cost of processing electronics is to improve economies of scale through increased volume. Until recently, the primary sources of electronic devices for recycling have been manufacturers and large businesses. [11] Redirecting household-generated e-waste, that has typically been disposed of in a landfill or left unused in storage, could provide recyclers with a larger and steadier supply of products to recycle.

According to the Government Accountability Office (GAO), cost and inconvenience inhibit consumers from recycling used electronics. [12] Although some computer manufacturers now accept their own products for recycling free of charge, [13] consumers generally have to pay the manufacturer a fee to recycle their e-waste (by packing and shipping them to the manufacturer themselves) or they must drop off their used electronics at often inconvenient locations (also, often for a fee). If the consumer does not pay for recycling, any recycling fees would likely be absorbed by the state or local agency collecting the device.

Most stakeholders agree that if e-waste is to be recycled, it must be as easy for consumers to recycle electronics as it is to buy them. [14] Many local and state agencies, retailers, and electronics manufacturers have worked with EPA to sponsor pilot programs providing convenient, free recycling services to consumers. The success of those programs demonstrated how successful e-waste recycling programs can be if they are convenient and inexpensive. [15] However, most states do not want to bear the full financial burden of establishing an e-waste management program. One factor driving states to develop e-waste laws is to implement a system that will provide financing for an e-waste collection, transportation, and recycling system.

The Inability of Stakeholders to Agree on a National System

For several years, interested stakeholders have debated how to best address the e-waste management issue. Those stakeholders include electronics manufacturers and retailers, local and state governments (particularly waste management and water treatment agencies), recyclers, environmental organizations, and charitable organizations that accept donation of used electronics. In general, these groups have agreed that the growth of e-waste has outpaced the development of infrastructure to appropriately reuse or recycle it. However, those stakeholders have disagreed about the best way to implement a program to manage such waste.

In 2001, a group of stakeholders formed the National Electronics Product Stewardship Initiative (NEPSI). The group's mission was the "development of a system, which includes a viable financing mechanism, to maximize the collection, reuse, and recycling of used electronics, while considering appropriate incentives to design products that facilitate source reduction, reuse and recycling; reduce toxicity; and increase recycled content." [16] Relatively early in the process, NEPSI determined that federal legislation would be required to implement any plan agreed to by the group. [17]

By 2004, the group had reached an impasse on how to finance a nationwide recycling system. That impasse divided the group into two camps — those who believed that a collection and recycling program should be financed through a consumer-paid advance recycling fee (ARF) assessed at the point-of-sale of designated electronic devices, and those who advocated a "producer pays" model wherein electronics manufacturers either took back their own e-waste and recycled it or paid for a system that would. (For more information about these two financing systems, see the section "A Mechanism to Fund the Program," below.)

In the absence of federal e-waste legislation or stakeholder consensus regarding an appropriate national e-waste collection and recycling program, states have begun to implement their own programs. Provisions of each law vary significantly and range from a ban on the landfill disposal of CRTs to implementation of a state-wide e-waste collection and recycling program.

Common Provisions of State E-Waste Laws

Twelve states have enacted some form of e-waste management law. Although the goals of each law are similar — to avoid landfill disposal of certain e-waste —the approaches taken to achieve those goals differ significantly. However, most state laws and proposals have certain broad elements in common, such as specifying the electronic devices covered under the law, how a collection and recycling program will be financed, collection and recycling criteria that must be implemented to minimize impacts on human health and the environment, and restrictions or requirements that products must meet to be lawfully sold in the state. [18]

Definition of Products Covered by the Law

Each state e-waste management program identifies certain electronic devices covered under the law. Most often, these are referred to as "covered electronic devices" (CEDs). Each state e-waste law defines CEDs slightly differently, but all include cathode ray tubes (CRTs). All state laws apply to CRTs in computer monitors, most also include CRTs in televisions. The laws also apply only to CRTs above a designated screen size — generally greater than four inches measured diagonally. CEDs may also include

- desktop computers — including the central processing unit (CPU);
- flat panel computer monitors or televisions using a plasma display or liquid crystal display (LCD);
- portable computers (laptops);
- combination units (CPUs with monitors);
- peripheral devices, such as keyboards, printers, and other devices sold for external use with a computer;
- facsimile (fax) machines;
- DVD and video cassette players or recorders; and
- cell phones. [19]

Each state law also specifies the types of electronic devices that are *not* regulated under the law, which usually includes video displays that are contained within a motor vehicle or piece of industrial, commercial, or medical equipment, and certain consumer products, such as clothes washers or microwave ovens.

A Mechanism to Fund the Program

In attempting to reach consensus among the various stakeholders, determining "who pays" for e-waste collection and recycling programs has been the most contentious issue. Many potential methods exist for funding an e-waste collection and recycling program. [20] Most current state programs fall into two broad categories: a consumer-paid system or a producer- or manufacturer-paid system. [21]

Under any financing scheme, it must be determined who will pay to manage orphan waste. "Orphan waste" is e-waste for which no manufacturer can be identified or for which the manufacturer is no longer in business. Given the high rate of turnover in the electronics business, the large numbers of foreign manufacturers that may be difficult to track down, and the large numbers of "white box" products (electronic products put together by component assemblers without a brand name affixed to the device), the proportion of orphan waste is potentially substantial. For example, as a part of its e-waste program, Washington State identified more than 1,200 orphan brands of electronics potentially sold or likely to appear in the state's e-waste stream. [22]

The Consumer Pays Model

One method of financing an e-waste collection and recycling program requires consumers to pay an advanced recycling fee (ARF) at the point-of-sale of designated electronics. Proceeds from the ARF would be used to implement the state's e-waste collection infrastructure and recycling programs. The collected funds may be managed by a state commission or a private third-party organization (TPO). Suggested ARFs are generally between $6 and $10, depending on the size of the device. This amount is less than the cost of recycling individual CEDs. However, because more products are sold than enter the waste stream, the cost of establishing a recycling infrastructure may be paid for at less than the per-unit recycling price. A portion of the funds may pay for local collection so that the government does not assume the cost of developing and running the recycling infrastructure. Also, retailers collecting the fee may keep a certain percentage of the fee to cover their administrative costs.

The ARF approach has been adopted in California and is favored by certain electronics manufacturers. Supporters of this system argue, in part, that it would

- immediately and reliably create a sustainable source of funding for a recycling infrastructure;
- pay for recycling all returned products, including orphan waste;
- be simple and could be implemented efficiently; and
- include a fee that is transparent to consumers and may contribute to consumer awareness of the need to recycle.

Stakeholders opposed to this approach argue, in part, that

- it would not be easily applied to Internet sales, presenting a competitive disadvantage to retailers assessing the fee;
- it would not provide manufacturers with an incentive to produce more environmentally benign products or to design products that may be more easily recycled; and
- if the collected funds exceeded recycling costs, those funds could be taken by the state and used for other purposes.

The Producer Pays Model

This approach requires manufacturers to implement or finance a collection and recycling program that takes responsibility for their share of returned e-waste and a share of orphan waste (generally assessed based on the company's total market share in the state). This is often referred to as an extended producer responsibility (EPR) or producer-pays approach. Some models allow the producer to address its share of waste by establishing its own recycling program (or one in cooperation with other manufacturers) or paying the state for the collection, consolidation, and recycling costs of its share. To date, all state e-waste programs, except California's, implement some version of the producer-pays model.

Stakeholders in favor of this approach include environmental organizations, retailers, and certain manufacturers that have already established recycling programs for their products. Some of their arguments to support this approach include

- the system places limited responsibilities on retailers and consumers and avoids the creation of new taxes on consumers,
- manufacturers implementing their own recycling programs have the flexibility to design their recycling program as they see fit, and
- making manufacturers responsible for recycling their own products may make them more likely to design products that are easier to recycle or that would have fewer toxic components.

Stakeholders opposed to this approach argue, in part, that

- the allocation of disposal/recycling costs to a given manufacturer would likely require costly sorting to identify the appropriate manufacturer,
- existing manufacturers would be responsible for the cost of recycling orphan waste, and
- the internalization of recycling costs may ultimately cost consumers more than an ARF once the recycling costs are subject to mark-up as the product moves through the distribution process.

Another variation on the producer-pays model involves manufacturers paying a flat fee to sell their products in that state. The collected fees are used to create a grant program for local governments to implement an e-waste recycling program. Maryland is currently the only state with such a program.

Collection and Recycling Criteria

State e-waste programs do not specify *how* e-waste must be collected. That is, they do not specifically require curbside pickup, municipal drop-off centers, retailer collection, or producer-established drop-off centers. However, each state program includes certain provisions regarding e-waste collection and recycling criteria that are intended to protect human health (particularly the health of individuals involved with recycling operations) and the environment.

Ban Landfill Disposal or Incineration

Although the details of state e-waste programs vary, one goal they all share is to reduce landfill disposal or incineration of e-waste, particularly CRTs. To reach that goal, some states have chosen to ban the disposal of CEDs in municipal solid waste landfills or incinerators.

Several states have implemented only a landfill or incineration ban (i.e., they have established no program to fund a collection and reuse/recycling program). Those states are Arkansas, Massachusetts, New Hampshire, and Rhode Island. In some states, a landfill ban preceded implementation of a full e-waste recycling program (see discussion regarding the

California and Minnesota programs, below). In other cases, implementation of an e-waste collection and recycling program is required before a landfill ban takes effect (see discussion regarding the Connecticut program, below).

No state specifically requires consumers to recycle. Therefore, if a landfill ban is in place, the responsibility generally falls solely on the municipal government to collect e-waste and ensure that it is properly managed (i.e., not sent to a municipal landfill).

Restrict E-Waste Exports

According to the Department of Commerce, much e-waste (possibly 50% to 80%) is sent overseas for recycling because it is more costly to recycle in the United States and most consumer electronics manufacturers (who provide the market for materials recovered from recycled electronics) have moved overseas. [23] Also, in states that ban the disposal of CRTs in landfills and incinerators within their borders, e-waste can be recycled, disposed of outside of the state, or exported. If a recycling infrastructure is not present before a disposal ban takes effect, exporting e-waste may be the most likely choice. Even if there are recyclers present in a given state, recyclers manage a significant percentage of the e-waste they receive by exporting it.

Most often the exported e-waste is sent to nations such as China or developing countries of Asia. Environmental organizations and certain other stakeholders are concerned that those countries do not enforce environmentally sound waste management practices or recycle in a manner that would protect workers handling toxic materials. [24] These practices potentially expose vulnerable populations to toxic chemicals, with few, if any, worker protections or a framework to protect the local environment. Some states have responded to these concerns by banning e-waste exports.

Set Recycling Standards

Some state programs address potential environmental concerns by attempting to ensure that the law does not exchange one potentially harmful disposal method (e.g., disposal in a solid waste landfill) for another (e.g., recycling in a manner that may harm employees or the environment). They may do so by directing their state environmental protection agency to develop recycling standards. Those standards generally specify criteria that should be met to ensure that e-waste is recycled in compliance with all applicable environmental, health, and safety regulations, and in a manner that protects the environment and the health and safety of workers in the United States and other countries.

Prohibit the Use of Prison Labor

Federal Prison Industries, a government-owned corporation that does business under the trade name UNICOR, runs e-waste recycling programs employing prison laborers. Some stakeholders are opposed to the use of UNICOR. For example, some recyclers have cited unfair competition from UNICOR, which they see as an impediment to creating a competitive recycling market because UNICOR's low labor rates keep prices down. [25] Others cite health and safety problems that have led to inmate workers being exposed to toxic and hazardous components. [26] These concerns have led some states to include a ban on the use of prison labor in their recycling programs.

Product Restrictions

Most state e-waste laws specify some conditions that manufacturers or retailers must meet before a product can be offered for sale in the state.

Labeling Requirements

Most state e-waste programs implemented to date require some entity (e.g., a state agency, e-waste collector, or other third party) to determine the share of collected e-waste that can be attributed to individual manufacturers. In order to more easily identify responsible manufacturers, most state laws specify that a manufacturer may not offer for sale in the state a CED unless it has a visible, permanent label clearly identifying the manufacturer of the device.

Registration Requirements

Most state e-waste programs require electronics manufacturers to register with the state. Generally, states require an initial registration, an annual registration thereafter, and payment of a registration fee. Information required to be included in the registration varies significantly from state to state. States may also require e-waste collectors, transporters, and recyclers to register with the state in order to be paid for their services.

Restrictions on the Use of Certain Materials

Although not common, state e-waste laws may include certain elements of European Union (EU) Directive 2002/95/EC on the restriction on the use of certain hazardous substances in electrical and electronic equipment (EEE). [27] Known commonly as the RoHS Directive, it bans the use of certain heavy metals and brominated flame retardants from EEE. The RoHS Directive requires that EEE does not contain lead, mercury, cadmium, hexavalent chromium, polybrominated biphenyls (PBBs), and polybrominated diphenyl ether (PBDE). Exemptions for certain applications of these substances are granted where substitution is not feasible or the potential negative environmental and/or health impacts caused by substitution outweigh the environmental benefits. By July 6, 2006, manufacturers selling electronic equipment in EU member states were required to have made the required substitution for hazardous substances.

Most electronics manufacturers sell to a worldwide market. Since they cannot easily change their production processes to accommodate different markets, it is likely that manufacturers that sell products in the United States will similarly meet the requirements of the RoHS Directive. For those manufacturers that do not, RoHS-like provisions in even a small number of states could have the effect of a nationwide requirement. To date, only California has included RoHS-like provisions in its e-waste law by prohibiting the sale of electronic devices that would be prohibited for sale under the RoHS Directive.

Retailer Restrictions

Some state laws make retailers a party to enforcing the law, in essence, by prohibiting them from selling CEDs that do not meet certain requirements. For example, retailers may be required to sell only those CEDs from manufacturers that are registered with the state or that meet labeling requirements.

OVERVIEW OF ENACTED STATE LEGISLATION

State e-waste laws are similar in that they intend to facilitate the recycling of certain electronic devices. Although the designated CEDs vary slightly from state to state, all include CRTs from computer monitors. Aside from that similarity, each state's means of achieving its goals is significantly different. Following this overview of enacted state e-waste laws, Table 1 compares key elements of each program.

Arkansas

On April 9, 2001, the state enacted the Arkansas Computer and Electronic Solid Waste Management Act. [28] The law applies to computer and electronic equipment —defined as a personal computer, computer component, audio player, stereo player, videocassette player, facsimile machine, copy machine, cellular telephone, wireless paging device, video game console, or any electronic item containing an intact or broken CRT. The law authorizes the state's Department of Environmental Quality to establish and implement rules and regulations banning the disposal of all computer and electronic equipment in Arkansas landfills by January 1, 2008 (that deadline was extended from the original date of January 1, 2005).

The law also establishes a program that requires state agencies to develop plans to sell, reuse, recycle, or dispose of surplus computer equipment and electronics; and encourages those agencies to donate unsold equipment to Arkansas public schools. A portion of the funds generated from selling surplus electronics must be allocated to a Computer and Electronics Recycling Fund, also established by the law. Among other activities, funds may be used for product market research and development grants to determine the most efficient means of collecting, transporting, and processing scrap electronic equipment, and to establish statewide contracts for computer and electronics recycling and demanufacturing businesses.

California

California's Electronic Waste Recycling Act of 2003 was enacted on September 24, 2003, and subsequently amended September 29, 2004. [29] The law applies to new or refurbished televisions or computer monitors that use a CRT or liquid crystal display (LCD), laptop computers, or any other video display device larger than four inches.

Beginning January 1, 2005, the law requires that at the time of retail sale, California consumers must pay an "electronic waste recycling fee" ranging from \$6 to \$10, depending on screen size. No recycling fee is assessed on the resale or reuse of a covered device. Retailers are required to transfer the collected fees to the Board of Equalization, which in turn deposits the money into an account managed by the California Integrated Waste Management Board (CIWMB). The CIWMB distributes the funds from this account to approved recyclers or to registered manufacturers that are collecting and recycling CEDs.

Following are additional requirements of California's e-waste program:

- A new or refurbished CED cannot be sold without a clearly visible label showing the name of the manufacturer or the manufacturer's brand.
- CED wastes cannot be exported to a foreign country without proper notification to the state Department of Toxic Substances Control.
- The sale of electronic devices that would be prohibited for sale under the RoHS Directive would be prohibited for sale in the state.
- State agencies purchasing or leasing covered electronic devices must require prospective bidders to certify that they comply with the law.

A ban on the landfill disposal of CRTs went into effect under regulations issued by the state in 2001.

Connecticut

On July 6, 2007, Connecticut enacted its e-waste recycling law, [30] which applies to desktop or personal computers, computer monitors, portable computers, CRT-based televisions and non-CRT-based televisions or any other similar or peripheral electronic device. Under this law, by January 2009, manufacturers will be required to participate in a program to implement and finance the collection, transportation, and recycling of certain electronic devices. The law requires manufacturers to register with the state Department of Environmental Protection and pay an annual fee that the state will use to administer the recycling program.

Also by January 2009, municipalities must provide for the collection of CEDs; waste CEDs must then be transported to and recycled by an approved recycler. In addition to the registration fee, manufacturers must pay reasonable costs of transportation and recycling CEDs attributed to them, and will be billed a pro rata market share for orphan devices.

Following are additional elements of Connecticut's e-waste program:

- By January 1, 2008, a manufacturer or retailer cannot sell a CED in the state unless it has permanently affixed, readily visible label with the manufacturer's brand.
- To be eligible to receive funds from the state, CED collectors, transporters, and recyclers must meet performance standards established by the state.
- Retailers must provide consumers with information on recycling.
- By January 1, 2009, CEDs collected through any state program can not be exported for disposal in a manner that poses a significant risk to the public health or to the environment.
- A landfill disposal ban will take effect in January 2011.

Maine

On April 22, 2004, Maine enacted the Act to Protect Public Health and the Environment by Providing for a System of Shared Responsibility for the Safe Collection and Recycling of Electronic Waste. [31] The law applies to waste televisions and computer monitors (CRTs and flat panel displays or similar video display devices with a screen greater than four inches measured diagonally). [32] The law implements a version of the producer-pays model that requires manufacturers to pay for the handling, transportation, and recycling of televisions and computer monitors.

By July 20, 2006, municipalities were required to ensure that waste televisions and computer monitors generated by households are recycled. They are required to ensure that a system is in place for delivering residential waste televisions and computer monitors to a consolidation facility in Maine. Each municipality may determine how this requirement will be met (e.g., operate an ongoing collection center, have one-day collections, or have residents deliver directly to a nearby consolidator).

Beginning January 1, 2006, consolidation facilities were responsible for counting each household-generated waste television and computer monitor and determining the total from each manufacturer. By March 1 of each year, beginning in 2007, the consolidator must provide this accounting to the state Department of Environmental Protection and submit a bill to manufacturers for allowable costs associated with recycling (i.e., the costs of handling, transportation, and recycling of their own television and computer monitor products, plus a pro rata share of orphan products). The consolidator must also transport waste televisions and computer monitors to a certified recycling and dismantling facility.

Following are additional elements of Maine's e-waste program:

- All CEDs offered for sale in the state must have a visible, permanently affixed label clearly identifying the manufacturer of the device.
- Retailers are prohibited from selling any CED offered for sale by a manufacturer not in compliance with the law.
- Recyclers must provide consolidators with a sworn statement that their operations meet environmentally sound management guidelines established by the state.

A ban on the landfill disposal of mercury-added products and CRTs went into effect under separate law on July 20, 2006.

Maryland

The law implementing Maryland's Electronic Recycling Program took effect on July 1, 2005, and was subsequently amended with changes that take effect October 1, 2007. [33] The law, implemented as a pilot program scheduled to end December 31, 2010, applies to manufacturers of computer or video display devices (CRT, LCD, plasma, digital, or other image-projection technology) with a screen greater than four inches.

To sell CEDs in Maryland, manufacturers of more than 1,000 devices a year must register with the Maryland Department of the Environment and pay a $10,000 registration fee (before the amendment, the fee was $5,000). In subsequent years, manufacturers must pay a $5,000 annual fee. Manufacturers with an approved takeback program pay an annual fee of $500. The fees are deposited in a fund to make grants to counties and municipalities to implement local recycling plans and address methods for the separate collection and recycling of CEDs.

Following are additional elements of Maryland's e-waste program:

- All CEDs offered for sale in the state must be labeled with the name of the manufacturer name or the manufacturer's brand label.
- Retailers are prohibited from selling any CED offered for sale by a manufacturer not registered with the state.
- The law specifies criteria a manufacturer must meet to demonstrate that it has implemented its own takeback program.

Massachusetts

A ban on the landfill disposal of CRTs from television and computer monitors has been in effect since April 1, 2000. [34] The law makes it illegal for a person to dispose of, or for a landfill, incinerator, or transfer station to accept, CRTs for disposal. As part of its electronics recycling strategy, the state has established a grant program providing free electronics recycling for municipalities.

Minnesota

On May 8, 2007, Minnesota enacted its e-waste recycling law, [35] which applies to computers, peripherals (keyboards, printers, or other devices sold for external use with a computer), facsimile machines, DVD players, video cassette recorders, and "video display devices" — defined as a television or computer monitor, including laptops, with a CRT or a flat panel screen that is larger than nine inches.

Beginning July 1, 2007, the law requires manufacturers to register with the state and pay an initial registration fee of either $5,000 (those that sell more than 100 units per year in the state) or $1,250 (those that sell under 100 units). Thereafter, manufacturers must pay $2,500, plus a variable recycling fee based on the total weight of CEDs sold in the previous year. In addition to the registration fee, manufacturers must pay for collection and recycling of their e-waste.

The law also requires manufacturers to meet specific recycling goals (Minnesota is the only state to set such mandatory goals). During the first program year (July 1, 2007 through June 30, 2008), manufacturers must collect and recycle an amount equal to 60% of the total weight of CEDs sold in the state in the previous year; this amount increases to 80% in subsequent program years. Starting August 1, 2008, to assist manufacturers in determining the total weight of CEDs sold in the state, retailers are required to report to manufacturers the number and type of video display devices sold to households in Minnesota during the program year.

Following are additional elements of Minnesota's e-waste program:

- All CEDs offered for sale in the state must be labeled with the manufacturer's name or brand label.
- Retailers are prohibited from selling any CED offered for sale by a manufacturer not registered with the state.

- In addition to manufacturers, CED recyclers and collectors must also register with the state (but pay no fee).
- By September 2008, manufacturers' annual registration must include a report on any CEDs they sell that do not meet RoHS requirements.
- Recyclers are prohibited from using prison labor.

A ban on the landfill disposal of CRTs went into effect under separate law on July 1, 2006. [36]

New Hampshire

On May 24, 2006, New Hampshire enacted a ban on the landfill disposal and incineration of "video display devices." [37] Video display devices are defined as a "visual display component of a television or a computer, whether separate or integrated with a computer central processing unit, and includes a cathode ray tube, liquid crystal display, gas plasma, digital light processing, or other image projection technology, greater than four inches when measured diagonally, and its case, interior wires, and circuitry." The ban took effect July 1, 2007.

North Carolina

On August 31, 2007, North Carolina enacted the Solid Waste Management Act of 2007. [38] Section 16 of the law includes provisions regarding the management of CEDs (referred to in this law as "discarded computer equipment"). CEDs are defined as any desktop CPU or laptop computer, the monitor or video display unit for a computer system, and the keyboard, mice, and other peripheral equipment (not including printers, scanners, or fax machines). The definition does not include televisions.

By January 1, 2009, manufacturers that sell more than 1,000 items of computer equipment per year must register with the state and pay a $10,000 registration fee and an annual renewal fee of $1,000. Manufacturers are prohibited from selling computer equipment in the state unless an item has a visible, permanent label identifying the manufacturer affixed to the equipment.

Within 120 days of registration, manufacturers will be required to submit to the state a plan for the reuse or recycling of equipment. Among other provisions, recycling plans must describe any take-back programs that will be implemented; how the manufacturer will implement and finance the plan; and how it will transport discarded equipment from collectors. [39] Manufacturers may implement recycling plans jointly with other manufacturers. Standards for recyclers are not specified in the law. However, manufacturer recycling plans are required to "provide for environmentally sound management practices to transport and recycle discarded computer equipment." Manufacturers must also submit an annual report to the state that includes an evaluation of the recycling rate for equipment. However, the law does not specify recycling goals that manufacturers must meet.

Also by January 1, 2009, the state Department of Environment and Natural Resources will be required to maintain a list of registered manufacturers; implement a public education program regarding computer equipment reuse and recycling; and provide technical assistance to local governments on the establishment and operation of equipment collection centers.

Oregon

On June 7, 2007, Oregon passed its e-waste recycling law, [40] which applies to televisions and computer monitors of any type with screens larger than four inches, and to desktop or portable computers. By January 1, 2009, manufacturers that sell these products in the state will be required to finance "free, convenient, and environmentally sound" recycling services. Manufacturers can create their own take-back program or participate in a common program, but they must pay for collection, transportation, and recycling costs.

Manufacturers must also register with the state and pay an annual registration fee of $20, $200, $5,000, or $15,000, depending upon their market share in the state. The registration must include a list of products sold in the state and a statement regarding whether the manufacturer will implement its own recycling program in accordance with criteria established by the law, or use the state contractor program.

Among other requirements, the state Department of Environmental Quality will be responsible for maintaining a list of registered manufacturers and orphan brands; determining each manufacturer's return share of CEDs; establishing a state contractor program to collect, transport, and recycle CEDs; and determining the recycling fee to be paid by each registered manufacturer. State and local governments will fund consumer education and promotion of the law. At the time of sale, retailers will be required to provide consumers with information about where and how they can recycle CEDs in the state.

Following are additional elements of Oregon's e-waste program:

- A manufacturer may not sell any CED in the state unless it is labeled with its brand.
- A retailer may not sell any CED unless it is properly labeled and is on the list of registered manufacturers posted by the state.
- A ban on the landfill disposal of CEDs becomes effective January 1, 2010; the state Department of Environmental Quality may postpone the prohibition in any area of the state where there is an inadequate system for CED collection, transportation, and recycling.

Rhode Island

On July 7, 2006, Rhode Island enacted the Electronic Waste Prevention, Reuse and Recycling Act. [41] The law bans the landfill disposal of desktop computers (including CPUs), computer monitors, including CRT monitors and flat panel monitors, laptops, combination units (CPUs with monitors), CRT- and non-CRT-based televisions (including plasma and LCD), or any similar video display device with a screen greater than four inches diagonally and that contains a circuit board. The law specifies that after July 1, 2008, no person shall dispose of a CED in any manner other than by recycling or disposal as hazardous waste.

The law also requires the state Department of Environmental Management, in consultation with stakeholders, to develop a plan for implementing and financing a program that addresses the collection, recycling, and reuse of covered electronic products from all covered electronic product generators in the state. Progress reports on the study were due to the general assembly on January 1 and May 1, 2007. By December 31, 2007, the law also requires the department to submit to the general assembly a plan and recommendations for any legislation necessary to implement the plan for collection, recycling, and reuse of CEDs.

Texas

On June 15, 2007, Texas enacted its e-waste recycling law, [42] which applies to computer equipment — defined as desktop or notebook computers, including computer monitors or other display devices that do not contain a tuner (i.e., it does *not* include televisions). The law requires manufacturers to implement a recovery plan that provides consumers with a free and convenient program to recycle the manufacturer's computer equipment.

Following are additional elements of Texas's e-waste program:

- A manufacturer may not sell any CED in the state unless it is labeled with its brand.
- Retailers may not sell any CED unless it is properly labeled and is on the list of registered manufacturers maintained by the state.
- The Texas Commission on Environmental Quality (the Commission) is required to adopt standards for recycling such as those provided by the Institute of Scrap Recycling Industries, Inc. (which bans the use of prison labor).
- The Commission is required to educate the public regarding the computer recycling programs, maintain program information on a website, enforce requirements for recycling computer equipment, and compile and issue an annual electronic report to the state legislature.

Washington

On November 11, 2006, Washington passed its e-waste recycling law, [43] which applies to CRTs or flat panel computer monitors or televisions with a screen size of more than four inches, and to desktop or laptop computers. The law requires CED manufacturers to finance and implement a program to collect, transport, and recycle waste CEDs. The program must be implemented in accordance with requirements specified in a "Standard Plan," implemented by the state, that will apply to all manufacturers. Individual manufacturers may opt to implement their own "Independent Plan," if it is approved by the Washington Department of Ecology.

Table 1. Comparison of Selected Elements of State E-Waste Laws

Legislative Element	AR	CA	CT	ME	MD	MA	MN	NH	NC	OR	RI	TX	WA
Funding Mechanism													
ARF		✓											
Producer pays			✓	✓			✓		✓	✓		✓	✓
Flat fee					✓								
Collection & Recycling Standards or Restrictions													
Landfill ban	✓	✓	✓	✓		✓	✓	✓		✓	✓		
Export restrictions		✓	✓										
Ban on prison labor							✓					✓	✓
Recycling standards			✓	✓								✓	✓
Mandatory recycling goals							✓						
Product Requirements or Restrictions													
Product label to identify manufacturer or brand		✓	✓	✓	✓		✓		✓	✓		✓	✓
Provisions tied to the RoHS Directive regarding the use of certain hazardous substances		✓					✓						

Source: Table created by the Congressional Research Service (CRS), based on a review of state e-waste laws. See text for explanation of terms.

By January 1, 2007, and annually thereafter, manufacturers must register with the state and pay an annual administrative fee. By January 1, 2008, and annually thereafter, manufacturers must pay their apportioned costs associated with the implementation of the Standard Plan.

Initially, there is no provision to address financing orphan waste. However, by April 1, 2010, the state's Department of Ecology must report to the state legislature regarding the amount of orphan products collected. If more than 10% of the total products collected are orphan products, the department must provide recommendations for reducing the amount of orphan products or alternative methods for financing the collection, transportation, and recycling of orphan products.

Following are additional elements of Washington's e-waste program:

- A manufacturer may not sell any CED in the state unless it is labeled with its brand.
- The state is required to establish performance standards for environmentally sound management of CED processors, including financial assurances to ensure proper closure of a facility that is consistent with specified environmental standards.
- Retailers must provide information to consumers describing where and how to recycle CEDs and locations for collection or return of products.

- Each collector, transporter, and recycler of CEDs must annually register with the state.
- No plan or program may include the use of federal or state prison labor for CED processing.

References

[1] The Solid Waste Disposal Act (SWDA), enacted by Congress in 1965, provided federal statutory provisions regarding solid waste disposal practices. RCRA was a 1976 amendment to SWDA. All subsequent amendments to SWDA, including the Hazardous and Solid Waste Amendments (HSWA, P.L. 98-616) of 1984 and the Federal Facilities Compliance Act (FFCA, P.L. 102-386) of 1992, are commonly referred to as RCRA.

[2] Under RCRA, hazardous waste generators are regulated in accordance with the amount of waste they generate each month. EPA regulations specify three hazardous waste generator categories: large quantity generators (LQG, generators of more the 1,000 kilograms of hazardous waste per month), small quantity generators (SQG, generators of between 100 and 1,000 kilograms of hazardous waste per month), and conditionally exempt small quantity generators (CESQGs, generators of less than 100 kilograms of hazardous waste per month). CESQGs are largely exempt from RCRA's hazardous waste management requirements, but may have some recordkeeping or reporting requirements. Also, a CESQG may be subject to more stringent requirements established by the state.

[3] For more information, see the International Association of Electronics Recyclers "About Electronics Recycling," Web page at [http://www.iaer.org/aboutrecycling.htm].

[4] In April 2007, EPA published the results of its study on "Electronics Waste Management in the United States." The electronic products covered in EPA's analysis are televisions, personal computers (desktops, laptops, and computer monitors), hard copy computer peripherals (including printers, scanners, and fax machines), computer mice, keyboards, and cell phones. EPA used two different approaches to gather its data; the results from each approach are available at [http://www.epa.gov/ epaoswer/ hazwaste/recycle/ecycling/ manage.htm].

[5] Under the Deficit Reduction Act of 2005 (P.L. 109-171), Congress established a deadline of February 17, 2009 for the transition from analog to digital television. (For more information, see CRS Report RL31260, *Digital Television: An Overview*, by Lennard Kruger.) During this transition, it is anticipated that many consumers will choose to buy new digital televisions, potentially adding to the e-waste waste stream as old televisions are discarded.

[6] CRTs are the vacuum tubes that make up the video display components of televisions and computer monitors.

[7] Lead is a toxic metal that can cause delayed neurological development in children and other adverse health effects in adults, including increased blood pressure, nephritis, and cerebro-vascular disease. For more information, see EPA's Final Rule, "Hazardous

Waste Management System; Modification of the Hazardous Waste Program; Cathode Ray Tubes," 71 *Fed. Reg.* 42927 July 28, 2006.

[8] *E-waste: Swiss E-waste Guid*e, "Hazardous Substances," May 31, 2006, available at [http://ewasteguide.info/hazardous_substances].

[9] Polybrominated diphenyl ethers (PBDEs) are the most commonly used brominated flame retardants (BFRs) that became a replacement for polychlorinated biphenyls (PCBs). A study conducted in the late 1990s found that levels of PBDEs in samples of human milk had increased exponentially, in contrast to contaminants such as PCBs, since the early 1970s (see The Environment Canada Web page "Brominated Flame Retardants," at [http://www.nwri.ca/research/brf-e.html]). Still, the extent to which PBDEs pose a threat to human health is unclear (see the Department of Health and Human Services, Agency for Toxic Substances and Disease Registry, ToxFAQs™ for PBDEs, available at [http://www.atsdr.cdc.gov/tfacts68-pbde.html]).

[10] Tests designed to simulate landfill conditions have revealed that CRTs and certain other electronic devices exceed regulatory limits for lead leachability. See research by Timothy G. Townsend et al., "Investigation of TCLP Leachability of Leaded CRT Glass," sponsored by the Florida Center for Solid and Hazardous Waste Management, available at [http://www.ees.ufl.edu/homepp/townsend/Research/CRT/default.asp]; and "Leaching of Hazardous Chemicals from Discarded Electronic Devices," sponsored by EPA, Regions 4 and 5, available at [http://www.ees.ufl.edu/homepp/townsend/Research/ElectronicLeaching/ default.asp].

[11] The International Association of Electronics Recyclers, "About Electronics Recycling," Web page at [http://www.iaer.org/aboutrecycling.htm].

[12] Statement of John B. Stephenson, Director, Government Accountability Office (GAO) Natural Resources and Environment, before the Subcommittee on Superfund and Waste Management, Committee on Environment and Public Works, U.S. Senate, "Electronic Waste: Observations on the Role of the Federal Government in Encouraging Recycling and Reuse," July 25, 2005, pp. 8-10.

[13] In June 2007, Dell announced that it would recycle any of its products free of charge. For more information, see the "Dell and the Environment" Web page at [http://www.dell.com/ content/topics/segtopic.aspx/dell_recycling?c=us and cs=19 and l=en and s=dhs].

[14] Department of Commerce, Office of Technology Policy, *Recycling Technology Products: An Overview of E-Waste Policy Issues*, July 2006, p. 20, available at [http://www.technology.gov/reports.htm].

[15] Statement of John B. Stephenson, Director, GAO, "Electronic Waste: Observations on the Role of the Federal Government in Encouraging Recycling and Reuse," pp. 8-10.

[16] See the National Electronics Product Stewardship Initiative Web page at [http://eerc.ra.utk.edu/clean/nepsi/default.htm].

[17] See *Solid Waste Digest*, "E-Waste Brings Together Industry, Activists, And Government — But Is Consensus Near?" November 2002, at [http://www.wasteinfo.com/news/stories/ archives/2002/11/NA/N02B11.htm].

[18] Some states have implemented effective voluntary e-waste recycling programs as part of their household hazardous waste management program. Voluntary recycling programs (i.e., those that are not financed by an electronics manufacturer or through the assessment of an advanced recycling fee) are not discussed in this report. However,

for information about such a program, see the Florida Department of Environmental Protection's "End-of-Life Electronics" website at [http://www.dep.state.fl.us/waste/categories/electronics/default.htm].

[19] States that have requirements applicable to cell phones generally address that category of e-waste in separate legislation. Cell phone-specific requirements are not discussed in this report.

[20] For a description of additional funding mechanisms, see Department of Commerce, Office of Technology Policy, *Recycling Technology Products: An Overview of E-Waste Policy Issues*, July 2006, p. 83, available at [http://www.technology.gov/reports.htm].

[21] In discussing e-waste management, the terms "producer" and "manufacturer" are used interchangeably with regard to electronics manufacturers.

[22] See the Washington State Department of Ecology, "Washington Recycles: Electronics, Establishing a Return Share List" Web page, under "2009 Return Share Information: Orphan List," at [http://www.ecy.wa.gov/programs/swfa/ eproductrecycle/ return Share.html].

[23] The Department of Commerce report, *Recycling Technology Products*, p. 27.

[24] Basel Action Network and Silicon Valley Toxics Coalition, *Exporting Harm: The High Tech Trashing of Asia*, February 25, 2002, available at [http://www.ban.org/E-waste/technotrashfinalcomp.pdf].

[25] The Department of Commerce report, *Recycling Technology Products*, p. 30.

[26] Center for Environmental Health, Prison Activist Resource Center, Silicon Valley Toxics Coalition, and Computer TakeBack Campaign, *Toxic Sweatshops: How UNICOR Recycling Harms Workers, Communities, the Environment, and the Recycling Industry*, October 2006, available at [http://www.computertakeback.com/docUploads/ToxicSweatshops.pdf].

[27] The definition of "electrical and electronic equipment" covered under the RoHS Directive is substantially broader than any e-waste legislation proposed in the United States. It applies to virtually any device, within ten broad product categories, including "IT and telecommunication equipment," that depends on electric currents or electromagnetic fields to work properly. The Directive is available at [http://europa.eu.int/eur-lex/pri/en/oj/dat/ 2003/l_037/l_03720030213en00190023.pdf].

[28] The March 18, 2005 amended version of this law is available online at [http://www.arkleg.state.ar.us/ftproot/acts/2005/public/act970.pdf].

[29] See the California Integrated Waste Management Board Web page "Electronic Product Management: Statutes, Regulations, and Related Issues," at [http://www.ciwmb.ca.gov/Electronics/RegIssues/].

[30] See the Connecticut General Assembly Web page at [http://www.cga.ct.gov/2007/ACT/PA/2007PA-00189-R00HB-07249-PA.htm].

[31] See the Maine Department of Environmental Protection "E-waste" Web page at [http://www.maine.gov/dep/rwm/ewaste/index.htm].

[32] CEDs also include computer central processing units (CPUs) that contain one or more circuit boards, but only for labeling purposes (i.e., CPUs are not included in the collection and recycling program). Therefore, discussion of CEDs in the section of this report regarding Maine refers only to waste televisions and computer monitors.

[33] See the Maryland Department of the Environment's "eCycling in MD" Web page at [http://www.mde.state.md.us/Programs/LandPrograms/Recycling/SpecialProjects/ecycl ing.asp].

[34] See the Massachusetts Department of Environmental Protection "Computer Monitor and TV Recycling Options for Residents and Small Businesses" Web page at [http://www.mass. gov/dep/recycle/reduce/crtrsbz.htm].

[35] See the Minnesota Pollution Control Agency's "Minnesota's Electronics Recycling Law" Web page at [http://www.pca.state.mn.us/oea/stewardship/electronics-law.cfm].

[36] For information about Minnesota's CRT landfill ban, see [http://www.pca.state.mn.us/ oea/ stewardship/crt-ban.cfm].

[37] See the New Hampshire General Court Web page at [http://www.gencourt.state.nh.us/ legislation/2006/HB1455.html].

[38] See the North Caroline General Assembly Web page at [http://www.ncleg.net/ gascripts/ BillLookUp/BillLookUp.pl?Session=2007 and BillID=S1492].

[39] Discarded computer equipment collectors may include a municipal or county government, nonprofit agency, or retailer that accepts such equipment from the public.

[40] See the Oregon Department of Environmental Quality "Electronics (E-Waste) Recycling Program" Web page at [http://www.deq.state.or.us/lq/electronics.htm].

[41] See the State of Rhode Island General Assembly Web page at [http://www. rilin.state.ri.us/ publiclaws/law06/law06365.htm].

[42] See the Texas Legislature Web page at [http://www.capitol.state.tx.us/tlodocs/80R/ billtext/html/HB02714F.htm].

[43] See the Washington State Department of Ecology, "Washington Recycles: Electronics, Establishing a Return Share List" Web page at [http://www.ecy.wa.gov/programs/swfa/ eproductrecycle/index.html].

In: E-Waste: Management, Types and Challenges
Editors: Yuan Chun Li and Banci Lian Wang
ISBN 978-1-61942-217-9

Chapter 8

ELECTRONIC WASTE TREATMENT BY THERMAL PLASMA

Imed Ghiloufi*

College of Sciences, Department of Physics, Al-Imam Muhammad Ibn Saud University, Riyadh, Kingdom of Saudi Arabia

ABSTRACT

Expansion in electrical and electronic equipment trade has led to significant increase in electronic waste which should be dealt with special priority due to its potential negative impact to the public health and the environment. Thermal plasma technology offers a very promising alternative of hazardous waste treatment for the near future. Plasma treatment is advantageous where landfill is difficult or impossible, where storage costs are encountered, or where stringent emission standards make combustion-based treatment expensive. This is usually the case for hazardous materials, in particular for e-wastes which contain heavy metals.

Plasma technology is an effective solution to treat the most toxic wastes like organohalogen wastes, contaminated hospital wastes, fly ashes, radioactive wastes, and e-wastes. Plasma torch technology has been developed to produce temperatures as high as 4000 ° C. At these temperatures, any substance is vitrified, i.e., reduced in volume and transformed into a glassy obsidian whose leachability is many times less than bottle glass. Thermal plasma technology takes the advantages to incinerate the combustible parts of wastes for volume reduction and to vitrify the noncombustible counter parts simultaneously into glassy slags with very low leaching rate. Furthermore, and compared with other waste treatment technologies, the principal advantages of plasma technology are the reduction of exhaust gas flow rates, in most cases higher volume reduction, installations with smaller footprints and lower investment costs for a given throughput, and faster start-up and shut-down times.

In this chapter, we will define and give the different types and common forms of plasma. We will also present an overview of plasma processes for waste treatment, and give a review of solid waste treatment by plasma. In section 5 we will present the technique of hydrogen recovery from the thermal plasma gasification of solid wastes.

* Tel: 00966509407164; Fax: 0096612591616; E-mail: ghiloufimed@yahoo.fr.

However in section 6, we will present the long- term behavior of vitreous or crystallized materials obtained by vitrification of solid wastes. In the last section we will present a study on the e-waste treatment by plasma.

1. INTRODUCTION

The production of electrical and electronic equipment is one of the fastest growing businesses in the world. The useful life of consumer electronic products is relatively short and decreasing as a result of rapid changes in equipment features and capabilities. Technological innovation and market expansion are accelerating the replacement of outdated equipment, leading to a significant increase in waste that induces a new environmental challenge. Concerns over environmental issues arising from these electrical and electronic products have led to a number of environmental regulations such as the waste of electrical and electronic equipment (WEEE) and restrictions on hazardous substances (RoHS) directives. These directives directly affect the electrical and electronic industries throughout the world [1].

Electronic waste is considered to be one of the priority streams in waste management. Challenges faced by WEEE management are not only consequences of growing quantities of waste, but also the complexity of WEEE [2-3]. Electronic waste is loosely applied to consumer and business electrical and electronic equipment that is broken, near the end of its useful life or unwanted. It is one of the most complex waste streams because of the wide variety of products from mechanical devices to highly integrated systems and the accelerating technological innovations. Certain components of some electronic products contain materials that render them hazardous. Incineration is normally used for destruction and volume reduction of waste [4]. The problem of incineration is the production of residues containing hazardous materials, such as chloride and heavy metals, which cause secondary environmental problem. For this reason alternative thermal treatment technology, such as thermal plasma, is required for sterilization, and for volume reduction of the fly ash resulting from the incineration of e-wastes.

Need for high temperature thermal processing in various applications such as solid waste destruction has led to a rapid growth in the use of plasmas and plasma-based technologies into industry [5-9]. Hazardous materials may be destroyed by the high-temperature plasma energy under the reducing or the oxidizing condition of plasma gas, which can enhance chemical reactions.

In the following section, a definition and different types of the plasma is offered, and the common forms of plasma are outlined. An overview of plasma processes for waste treatment and a classification of waste materials follow in section 3. This section is followed by a review of solid waste treatment by plasma. Section 5 is devoted to the technique of hydrogen recovery from the thermal plasma gasification of solid wastes. The long- term behavior of vitreous or crystallized materials obtained by vitrification of solid wastes is briefly reviewed in section 6. In section 7, and before the conclusion, we present a study on the e-waste treatment by plasma.

2. Definition and Types of Plasma

2.1. Definition of Plasma

Plasma is the most common form of matter; it is in the stars and in the tenuous space between them makes up over 99% of the visible universe. In physics and chemistry, plasma is a state of matter similar to gas in which a certain portion of the particles are ionized. Some scientists have dubbed plasma the "fourth state of matter" because plasma is neither gas nor liquid, but its properties are similar to those of both gases and liquids.

When a gas is heated to some temperature its molecular bonds are dissociated and its molecules are rendering to atoms. However a significant increase in the temperature of this gas leads to its ionization (a loss of electrons), and it will contain a charged particles (positive ions and negative electrons). This ionized gas can be called plasma if it satisfies the following criteria:

1. Quasi-neutrality. A plasma maintains almost perfect charge balance:

   $$-q_e.n_e = q_i.n_i \pm \Delta$$

   where Δ is tiny. q_e, q_i, are the charges of electron and ion, respectively, whereas n_e, and n_i are the mole numbers of electrons and ions, respectively.
2. Interactions between individual charged particles are insignificant compared to collective effects. This introduces the concept of *Debye screening*. This condition requires that the number of particles in the *Debye sphere* Λ □ □ 1. Λ is also called the *plasma parameter*.
3. The electron-neutral collision cross-section is much smaller than the electron-ion cross section $\sigma_{en} << \sigma_{ei}$. Thus a weakly ionized (~ 1%) gas can behave like a plasma because the long range Coulomb forces have an effect much more significant than that associated with collisions with neutrals.

2.2. Common Form of Plasma

The common form of plasma and its applications can be divided in two classes. The first class is the industrial plasma, whereas the second concern the space and astrophysical plasmas.

a. Industrial plasma
   - Those found in plasma displays, including TVs
   - Inside fluorescent lamps (low energy lighting), neon signs
   - Rocket exhaust and ion thrusters
   - Inside a corona discharge ozone generator
   - Fusion energy research
   - Arcs produced by Tesla coils (resonant air core transformer or disruptor coil that produces arcs similar to lightning but with alternating current rather than static electricity)

- Plasmas used in semiconductor device fabrication including: Reactive Ion Etching, Sputtering, surface cleaning and Plasma Enhanced Chemical Vapor Deposition
- Laser-produced plasmas (LPP), found when high power lasers interact with materials.
- Inductively Coupled Plasmas (ICP), formed typically in argon gas for Optical Emission Spectroscopy or Mass Spectroscopy
- Magnetically Induced Plasmas (MIP), typically produced using microwaves as a resonant coupling method

b. Space and Astrophysical plasmas
- The Sun and other stars (plasmas heated by nuclear fusion)
- The solar wind
- The interplanetary medium (space between planets)
- The interstellar medium (space between star systems)
- The Intergalactic medium (space between galaxies)
- The ionosphere
- The polar aurora

2.3. Types of Industrial Plasma

There are two types of industrial plasma: cold plasma and thermal plasma.

a. Cold Plasma

Cold plasma or non-equilibrium plasma is a partly ionized gas, which is generated in a high-voltage electric field in a low pressure. It also characterized by high electron temperatures and low ion and neutral temperatures. The surface structure and properties of polymers, fibres and fabrics can be modified by cold plasma. Low temperature plasmas (cold plasma), used in surface modification and organic cleaning, are ionized gases generated at pressures between 0.1 and 2 torr. These types of plasmas work within a vacuum chamber where atmospheric gases have been evacuated typically below 0.1 torr.

b. Thermal Plasma

Thermal plasmas are atmospheric-pressure plasmas characterized by high enthalpy content and temperatures about 2,000-20,000 K. Thermal plasma device or plasma torch produces arc plasma column between two electrodes. There are several types of plasma torches: direct current (DC) arc torch, induction torch and high frequency capacitive torch [10, 11].

3. Waste Treatment by Plasma

Increasing population, consumerism and industrial development have led to an increase in the quantities of hazardous and municipal solid waste (MSW) generated worldwide. Although landfill continues to account for the majority of municipal solid wastes disposal, many countries and regions have limited landfill capacity and it is increasingly difficult to find new sites. For this reason various thermal processes, including incineration, pyrolysis, melting or vitrification, have been proposed for treating these hazardous wastes prior to disposal; their aim being to destroy the organic fraction and convert the inorganic fraction into an inert silicate slag, or glass, that can either be advantageously reused, or harmlessly disposed of in an inert landfill.

Use of plasma for waste treatment has had a fascination for many years because of the ability of the plasma to vaporize anything and destroy any chemical bonds. The principal advantages that plasma offers to treatment processes are [12, 13]:

1. The high-energy density and temperatures associated with thermal plasmas, and the correspondingly fast reaction times, offer the potential for a large throughput with a small reactor footprint.
2. The steep thermal gradients in the reactor permit species exiting it to be quenched at very rates so allowing the attainment of meta-stable states and non-equilibrium compositions, thereby minimising the reformation of persistent organic pollutants (POPs).
3. Plasmas can be used for the treatment of a wide range of wastes including liquids, solids and gases.
4. The high heat flux densities at the reactor boundaries lead to fast attainment of steady state conditions. This allows rapid start-up and shutdown times, compared with other thermal treatments such as incineration, without compromising refractory performance.
5. Oxidants are not required to produce the process heat source, as no fuel is combusted, therefore, the gas stream volume produced is much smaller than with conventional combustion processes and so is easier and less expensive to manage.
6. The combination of the above characteristics allows plasma treatment to be integrated into a process generating hazardous.

In addition, increased process controllability and flexibility, and smaller installation size, allow integration of the waste treatment process into the manufacturing process. The disadvantages associated with the plasma process, besides the already mentioned (largely perceived) lack of reliability, lie in the use of electricity, the most expensive form of energy. Consequently, economic considerations provide the strongest barrier for use of plasmas for waste treatment.

3.1. Overview of Approaches for Plasma Treatment of Different Types of Wastes

Plasma was used to treat several types of wastes such as:

- Destruction of chemical wastes and more particularly organohalogen wastes [14],
- Incineration of special wastes such certain contaminated hospital wastes [15],
- Vitrification of fly ashes [16],
- Vitrification of the grounds contaminated by in-situ action [17],
- Inertage of very refractory wastes containing asbestos (thermal insulators, joints high temperatures, etc. [16]).

Plasma treatment can consist of the following approaches [18].

a. Plasma pyrolysis, i.e. thermal breakdown of chemical components without oxidation.
b. Plasma gasification, i.e. incomplete oxidation of organic components of the waste and generation of a combustible gas, typically a mixture of carbon monoxide with hydrogen (syn-gas) with possibly some other gases included; this gas can be used for generation of hydrogen, for combustion in gas engines to generate electric power, or for heating steam to drive a steam turbine.
c. Plasma compaction and vitrification of solid wastes by gasifying organic material, melting inorganic material eliminating voids and binding hazardous metals in a ceramic matrix (e.g. a silicate) by adding appropriate fluxes; high plasma temperatures allow melting of high temperature fluxes resulting in vitrified ceramics with low leach rates.
d. Combinations of (a) and (c) or of (b) and (c), particular for solids with high fractions of organics.

3.2. Some Examples of Wastes Treated by Plasma

Although landfill continues to account for the majority of municipal solid wastes disposal, many countries and regions have limited landfill capacity and it is increasingly difficult to find new sites [19-21]. For this reason incineration of municipal solid waste (MSW) has become much more common, in particular in countries where use of landfills is tightly controlled. In Japan, about 70% of the municipal waste is incinerated, and 15–30% of the waste is collected as bottom ash (below the grate) or fly ash (in the filters) [22]. Most hazardous materials are usually contained in the fly ash because most toxic heavy metals easily evaporate, and fly ash treatment is also required in Europe. For an incinerator treating about 110 to 130 kt yr^{-1}, about 40 kt yr^{-1} of ash must be treated [23, 24]. For treatment of these incinerator residues, plasma technology offers distinct advantages. This is one of the most widespread uses of plasma waste treatment technologies. Besides resulting in a volume reduction of typically 50%, the toxic heavy metals can be encapsulated in a non-leaching glassy slag formed with the addition of high temperature fluxing agents with typical slag temperatures between 1500 and 1700 ◦C [18].

The solid wastes generated from the steel industry mainly consist of, zinc, lead, cadmium, stable oxides (such as calcia, silica and alumina), oxides of iron, chromium, nickel, manganese and phosphorus [25]. There are several possibilities for the treatment of these wastes including vitrification into a glassy slag, recovery of metals by oxide reduction, or a combination of both. Plasma can be successfully employed for all of these processes. The reduction of the oxides can be performed with a reducing plasma gas or with reducing additives (e.g. carbon). Metals such as zinc and lead can be collected as a melt or condensed from the vapor phase. The high temperature of the plasma also allows vitrification of the residuals after the recycling step [26].

For waste treatment, it is better to combine the plasma technology with another thermal treatment technology, plasma heating combined with either incineration or resistive heating should be emphasized. The first approach is particularly interesting when combustible exhaust gases are generated that are used for on-site generation of the required electric power.

In the second approach, the waste is resistively heated by a current, thus using electrical energy more efficiently for the melting of the material, while the plasma provides the required high temperature environment. A combination of conventional gasification technology with plasma treatment of residues is pursued by Tetronics [18]. For materials with a high fraction of metal content, combination of plasma treatment with induction heating may be of interest [27].

Incineration is not an effective solution for treating chloride compounds because dioxins and furans, produced by incomplete oxidation, remain in the off-gases as thermally persistent species. Undesirably, the waste heat recovery zone is an ideal location for the reformation of dioxins and furans. Pyrolysis followed by gasification by thermal plasma is a very attractive solution for these kinds of wastes because of the possibility of incorporating a dwell of at least 4 s at 1200 ◦C and rapid cooling from this temperature to minimize the possibility of forming dioxins or furans [28-30].

Plasma treatment of hospital solid waste allows a reduction of the exhaust gases to as low as 10% compared with incineration [31, 32], the possibility of having small local treatment plants and a volume reduction of the waste by 90%, which is larger than that achieved in incinerators. These types of waste have a relatively low negative value, i.e. any treatment must be inexpensive or favoured by regulations. A purely plasma based process for such wastes must offer significant advantages such as production of fuel gas for electricity generation.

Radioactive wastes are the most studied and supervised wastes since the explosions of Hiroshima and Nagasaki. This interest increased even further following the accidents of Three Miles Island and Tchernobyl. Radioactive wastes are classified according to the type of emitted radiations [33]. Depending on its category, radioactive wastes can be treated through either, storing on the surface, coating in various binders, or conditioning in a vitreous matrix by plasma vitrification. This latter treatment is reserved for the most toxic radioactive wastes such as fission products.

Moreover, several countries have started using plasma technology to treat radioactive wastes. For example, in Russia, the company RADON (for Moscow Scientific & Industrial Association) developed a plasma furnace to reduce the volume of weak activity wastes [34]. In South Korea, several companies joined to try out processes of conditioning weak and average activity wastes [35]. In Switzerland, the ZWILAG company (Zwischenlager Würenlingen AG) installed a process for plasma conditioning of waste resulting from nuclear centers, industrial laboratories, and hospitals [36]. Plasma is also used in Taiwan to treat weak activity wastes [37]. In Japan, several companies like the Central Research Institute of Electric Power Industry (CRIEPI), Japan Atomic Energy Research Institute (JAERI), and Japan Atomic Power Company (JAPC), using plasma furnaces for the conditioning and the reduction of the volume of weak activity wastes [38- 40]. In France and on the Cadarache and Marcoule sites, CEA (Commissariat of Atomic Energy) developed a reactor based on a plasma furnace for the incineration and the vitrification of radioactive wastes [33, 41].

4. Plasma Compaction of Solid Wastes

4.1. Description of Plasma Generating Devices for Waste Processing

A typical plasma system for treatment of solid wastes consists of (a) the plasma reactor, with collection of the metal (if applicable) and the slag at the bottom, periodically tapped and cast into some usable form, with power supply, cooling water supply gas supplies and control and data acquisition equipment, (b) a secondary combustion chamber for allowing sufficient residence time at elevated temperatures to assure complete reactions and gasification of soot; this secondary combustion chamber can be fired either by a burner or by a low power non-transferred plasma torch; (c) depending on the waste, a quenching chamber (usually water quench) to avoid formation of dioxins and furans, (d) a cyclone for particulate removal, (e) a scrubber for eliminating acidic gases, (f) if necessary a hydrogen sulfide absorber, (g) high efficiency filters or precipitators for small particulate removal, (h) an activated carbon filter for removal of heavy metals and (i) finally a fan for generating subatmospheric pressures in the entire installation. Additionally, various forms of waste preparation and feeding systems have to be integrated with the reactor. Most thermal plasmas are generated by either an electric arc or by a radio-frequency induction (rfi) discharge. In waste treatment, arc plasmas dominate because they are relatively insensitive to changes in process conditions [18].

The arc-generated plasma is used in two configurations: a transferred arc where one of the electrodes is usually the material to be treated and a non-transferred arc plasma, where the arc is contained inside a plasma torch and the plasma jet exiting the torch is used for processing [42]. In the transferred arc configuration, the waste material is exposed to the arc plasma which typically has peak temperatures of 12 000 to over 20 000 K, depending on plasma gas, location in the arc and the cooling of the arc. Minimal plasma gas flow rates are required for the operation of such an arc, however, high flow rates can be imposed. Transferred arc reactors exist for a very wide range of power levels, ranging from a few hundred watts to tens of megawatts. Transferred arcs can have multiple rod electrodes. Such an arrangement allows (a) the arcing between two rod electrodes for initial indirect heating of the waste, (b) the establishment of multiple series arcs between the electrodes and the waste material and (c) operation in ac or three-phase mode.

In contrast, the non-transferred arc plasma torch provides a plasma flow for treating the waste. The plasma exiting the water-cooled plasma torch has peak temperatures of 10 000–14 000K depending on the torch power level, the plasma gas and the torch design. However, since the plasma is decaying outside the torch, the material to be treated is usually not exposed to these high temperatures, and 'average' values are used for temperature and velocity of the plasma flow, i.e. values based on the mass averaged enthalpy flow.

4.2. Non-transferred Arc Reactors

In reactors with non-transferred arc torches, the plasma provides heat and the possibility of adjusting the chemical environment for generating useful off-gases. Two types of reactors are in use: the plasma fired shaft furnace developed by Westinghouse and the reactor in which the waste is heated directly by plasma jets, developed by Europlasma. While the fundamental developments have been performed in the 1980ies and 1990ies, the implementation for actual waste processing on a commercial scale has in general occurred in the last 8–16 years.

The Westinghouse approach is best illustrated by the plant designed by Hitachi Metals Corp. and operated in Utashinai, Japan [43, 44]. The plant has two furnaces and was designed predominantly for compaction of automobile shredder residue (ASR), with a 165 t d^{-1} capacity, but can also treat up to 300 t d^{-1} MSW. The installation contains four 300kW plasma torches heat air for a shaft furnace containing the waste and coke (about 4% of the total charge). The plasma heating of a fraction of the air reduces the amount of coke and air needed to generate the high temperatures in the furnace. The fraction of the air that is plasma heated is determined by the waste composition. Addition of calcinations agents allows production of ceramic tiles for construction. The off-gases contain a high fraction of carbon monoxide and hydrogen.

Burning these off-gases in a second reaction chamber provides the heat for steam generation which in turn is used to drive a steam turbine. Electricity is generated at a rate of 8MW, and less than half of the generated electric power is used for plant operation, the remainder is transferred to the electric power grid. Electric power requirements for the waste treatment range from 100 to 250 kWh t^{-1}.

Europlasma in France is using non-transferred arc torches in plasma reactors for incinerator residue compaction [45, 46]. Figure 1 shows a schematic of the installation in Cenon, France. The incinerator processes about 120 000 t yr^{-1} of MSW, and the plasma reactor processes up to 10 t d^{-1} of fly ash using a 500kW torch operating with air. The energy requirement is 1300 kWh t^{-1} of fly ash. Several larger installations exist in Japan using Europlasma designed reactors, most notably the plant in Shimonoseki, designed to process 42 t d^{-1} fly ash and bottom ash with a two torch reactor, each rated at 1.3MW. The specific energy consumption is about 800 kWh t^{-1}. All these reactors have been operating for several years with regular maintenance requirements of about 1 h per 300 h of operation. Europlasma has also developed a process for treating asbestos [47, 48]. The Inertam plant in St. Denis, France, heats the crushed asbestos-containing material with three plasma torches providing heat at different locations (see figure 2). One of the torches is rated at 500kW and the other two are rated at 2MWeach. The specific energy requirement is 1300 kWh t^{-1}. The material is fed continuously into the reactor at a rate of 20 t d^{-1}, and the slag is removed periodically. The vitrified product is being used in road construction. A post-combustion chamber is used as in many of the Europlasma installations to assure complete combustion of carbon-containing gases and to satisfy incinerator regulations.

A special situation is addressed in a development by Pyrogenesis Corp, namely that of compaction of shipboard waste [49–51]. With tighter regulation for ocean water pollution, volume reduction of shipboard waste has become an important issue. Plasma reactors offer the possibility of requiring minimal volume and small footprints. In the Pyrogenesis design (see figure 3) [51], the waste is pretreated by separating the metallic content, and shredding and milling the combustible material into a fine lint-like structure, and this material is injected into a plasma jet emanating from a non-transferred plasma torch. The mixing with the plasma occurs in a water-cooled 'eductor' shell (see figure 4) [51], and the resulting mixture of incompletely reacted gases is completely reacted with additional air in a secondary combustion chamber. Subsequent quenching with water minimizes the formation of dioxins and furans; 200 kg h^{-1} can be treated. Use of a plasma torch allows a rapid start-up time of about 5 min (since there are no refractory furnace walls to heat) and rapid shut-down times, minimization of the reactor size and lowering of the gas flow rates through preheating the combustion air. No fuel gases are needed. The torch is operating at about 109kW and provides about 15% of the energy needed for the treatment, the rest of the energy is provided by the heating value of the waste. The specific energy consumption, not counting the waste preparation, is about 665 kWh t^{-1}. The system has been installed in a cruise liner and has been operating for more than 16 years.

Phoenix Solutions Co. (Minneapolis, MN) has developed a reactor fired by a non-transferred plasma torch for processing of solids contaminated with PCBs [52]. The plasma torch operates at 150kW with nitrogen as plasma gas, and lamp ballasts have been fed into the reactor (see figure 5). The off-gases are quenched with water to avoid formation of toxic co-products. Extensive instrumentation has shown that contamination of the off-gas, the waste water, the waste water filter cake, as well as the slag leach rate are all well below the regulatory limits. Off-gases have a composition of about 30% hydrogen, 20% carbon monoxide and nitrogen as the plasma gas making up the majority of the remainder. Of particular interest is that the reactor design has been optimized by using advanced CFD simulation of the fluid flow and the temperature—time history of every fluid particle injected into the reactor or evaporated from the reactor wall. A similar reactor was developed for processing contaminated soil, with a 1 t h^{-1} feed rate and operating with a 1MW non-transferred plasma torch [22].

A pilot plant reactor for treating wastes with high chloride and fluoride content (i.e. plastics) is described by Bendix and Hebecker [53]. The waste is injected into a vertical shaft furnace and into a plasma generated by three plasma torches mounted on the walls of the reactor and running at 4–15kW each. Gas cooled silicon carbide walls allow operation at high wall temperatures. The plasma gas is a mixture of air, steam and part of the recycled exhaust gas. For treatment rates of about 20 kg h^{-1} a specific energy consumption of 1 kWh kg^{-1} is obtained. Performance is strongly dependent on optimization of the plasma waste mixing, and electrode life is so far limited to 50 h.

Rutberg [54] describes a system for treatment of medical waste based on the three-phase non-transferred plasma torch. Two plasma torches are used, one for treatment of the waste by firing arc heated air into a rotary kiln reactor, and the second for providing hot air for the secondary combustion chamber (thermal residence chamber). Throughputs of 200–300 kg h^{-1} are given, and a specific energy requirement on 0.8–1 kWh kg^{-1}.

Another reactor designed to treat fluorocarbon waste, including CF4 and Teflon has been described by Ponelis and v.d. Walt [55]. The material is introduced into a reactor fired by a 150kW non-transferred plasma torch at a rate of 10 kg h^{-1}. Quenching at a water-cooled cold surface allows recovery of some of the fluoride as C_2F_4 or C_3F_6 with a 40% yield. No specific energy requirement data are available.

A process similar to the Westinghouse–Hitachi approach and specifically designed for generation of syn-gas from organic wastes (with no halide content) is described by Rutberg *et al* [56] and by Bratsev *et al* [57]. A shaft furnace packed with waste is fired by a non-transferred plasma torch either from the top or from the side. Feed rates are in the order of 50 kg h^{-1}, and air and/or steam addition (depending on the waste) is used to generate the syn-gas which is extracted at the bottom of the reactor. Specific energy consumption is given as 1–2 kWh kg^{-1} depending on the waste material, and some excess electricity could be generated if the heating value of the syn-gas is converted to electricity with a 50% efficiency.

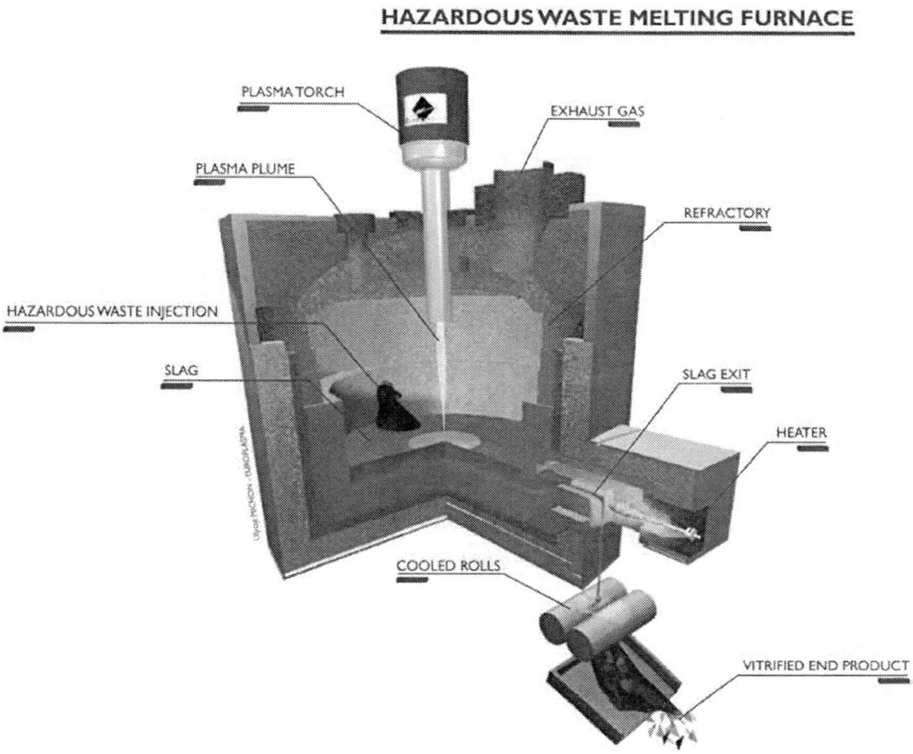

Figure 1. Schematic of Europlasma reactor in CENON for fly ash vitrification.

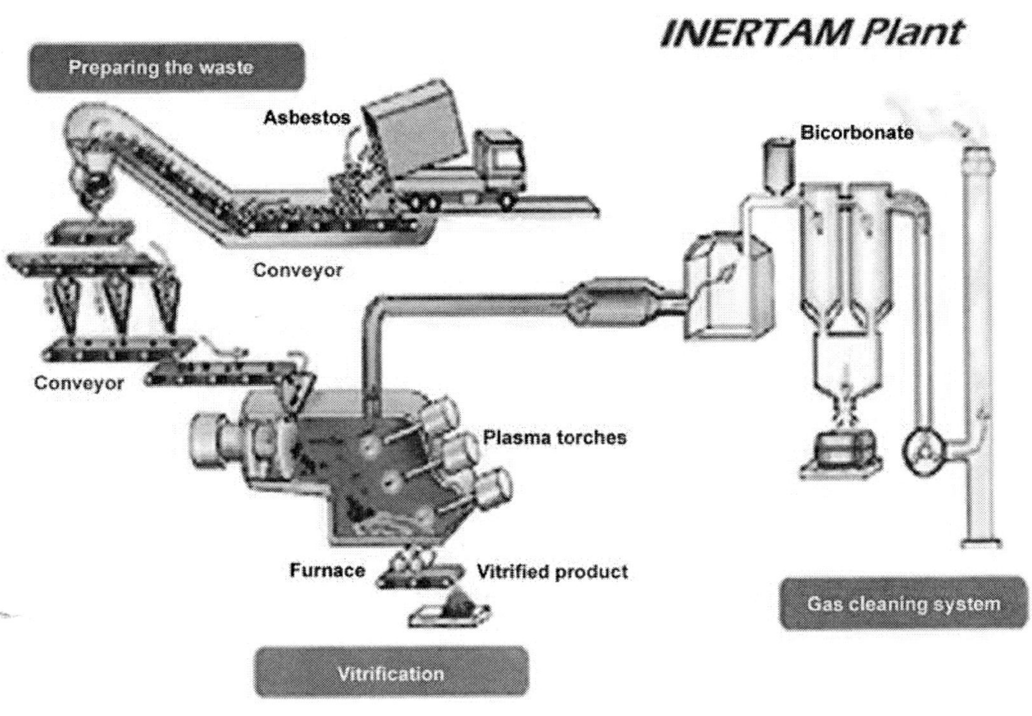

Figure 2. Schematic of the Europlasma asbestos treatment plant INERTAM with three Aerospatiale plasma torches.

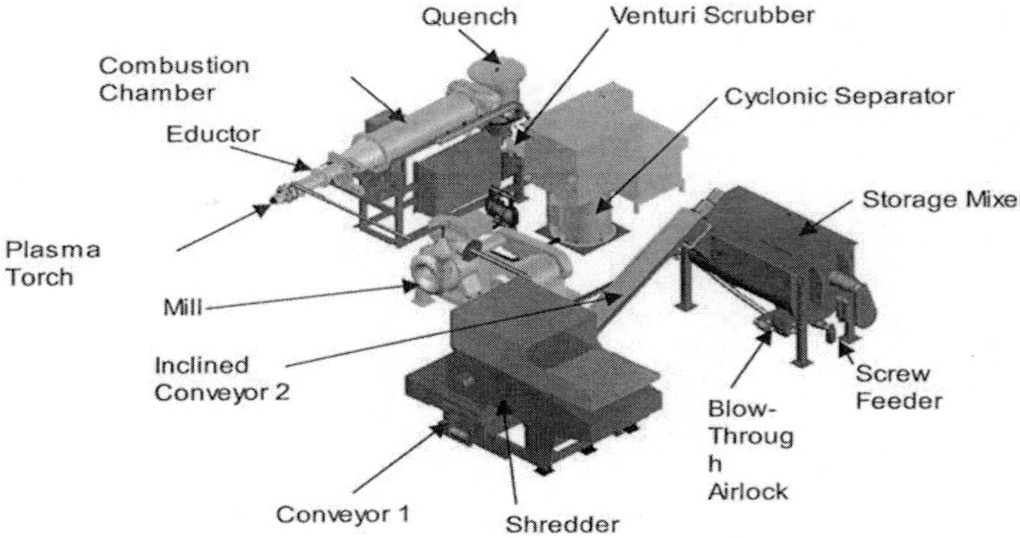

Figure 3. Schematic of Pyrogenesis shipboard waste treatment installation.

A process developed by HydroQuebec and commercialized by PASO (St Laurent, Canada) concentrates on the treatment of sludge from pulp and paper production [58]. A 200kW torch operating with air treats about 2 t h^{-1} of sludge in a rotary kiln. Most of the energy for drying the sludge is derived from reacting the organic material, and the plasma provides only about 10% of the required energy, but is necessary for a continuous reaction because of the high concentration of steam that is produced in the reactor. The steam, at temperatures above 500 °C, can be used for generating electricity.

4.3. Transferred Arc Reactors

There are numerous transferred arc reactor designs available for waste treatment, several of them based on successful developments for metallurgical plasma processes.

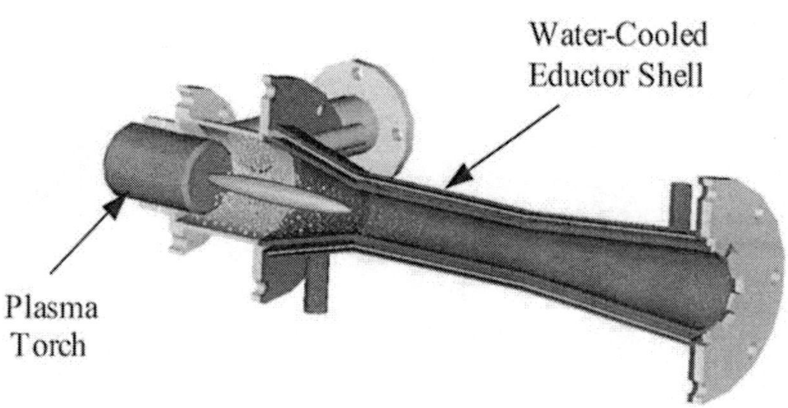

Figure 4. Pyrogenesis plasma fired eductor for gasifying ship board waste.

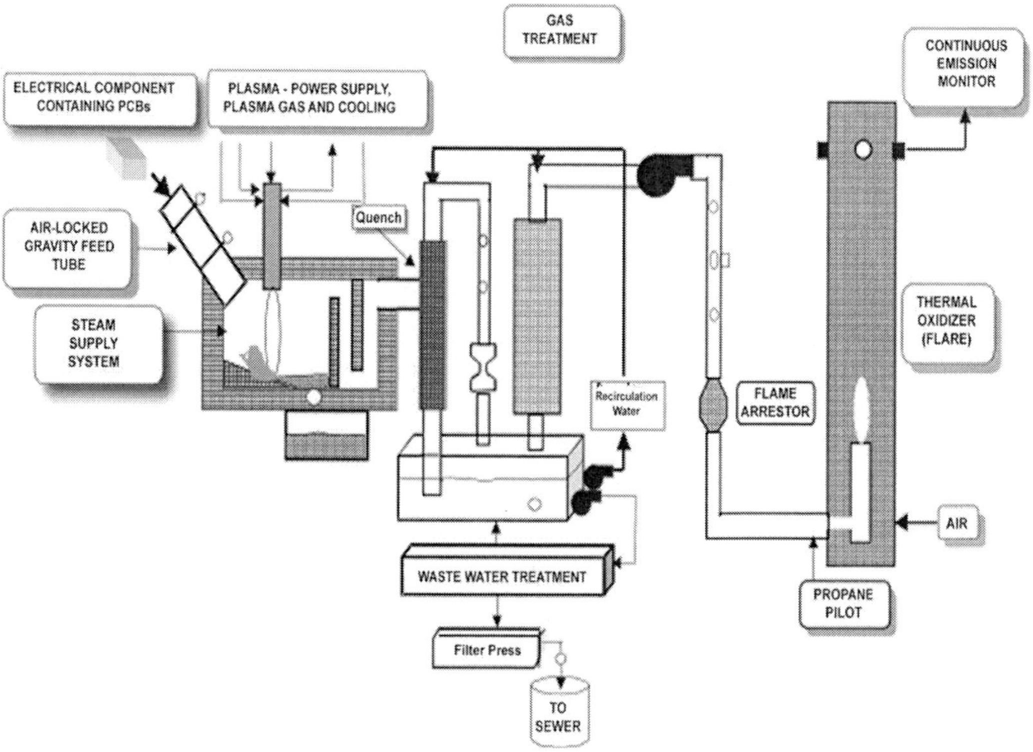

Figure 5. Schematic of Phoenix Solutions installation for processing contaminated solids.

Tetronics (Faringdon, UK) offers two types of reactors, single-torch reactors and twin-torch reactors [23, 60, 31, 61]. In the single-torch reactor, the arc is transferred from one electrode to the melt, the electrode is usually the cathode, and can be either water-cooled metal or graphite. Plasma gases are usually argon or nitrogen. The dominant single torch application is incinerator ash vitrification, and numerous installations are in operation, the majority in Japan. Operating powers range from 200 to 1200kW for metal cathode torches and up to 2.75MW for high current graphite cathodes (up to 9800 A). Corresponding treatment rates range from 200 kg h−1 to over 5800 kg h−1. Figure 6 [61] shows a schematic of one of these single-torch transferred arc reactors. A schematic of the Tetronics ash vitrification installation is shown in figure 7 [24].

Tetronics also offers a twin-torch reactor, where one of the torches serves as the anode while the other serves as the cathode (see figure 8) [62]. Again, the electrodes can be either graphite or water-cooled metal. In the case of the metal electrodes, a small amount of argon plasma gas flowing around the electrodes shields them from exposure to the gases from the waste, resulting in a typical electrode life of 200–300 h. The process gas is introduced through a shroud surrounding the electrode. In the case of graphite electrodes, the plasma gas is introduced through a hole down the centre of the electrode. Waste material is introduced into the plasma between the two electrodes. The twin-arc arrangement can be operated in the non-transferred mode, where the arc current flow is directly from one electrode to the other, or in the transferred mode, where the current flow is from one electrode to the charge, through the charge and then back to the other electrode. In this transferred mode case, the flow of the current through the melt provides Joule heating of the molten waste. The desired mode of operation is assured by changing the positions of the electrodes. Usually, the arc is initiated directly between the two electrodes until the reactor and the charge is at a sufficiently high temperature for the charge to be conducting. The furnace has water-cooled refractory lined side walls and roof, and a water-cooled copper crucible for containing the waste. During start-up, a small amount of waste is melted to establish a skull on the copper surface. Since the charge is not an electrode, electro-chemical reactions between the slag and the refractory liner are avoided.

A 250kW twin-torch reactor has been used for processing of 50 kg h^{-1} of bottom ash, with a specific energy requirement of 550 kWh t^{-1} [60]. Extensive leach tests show that heavy metal leach rates are well below the regulatory limits [23, 60]. Higher power installations are in operation up to 2800kW, processing up to 3300 kg h^{-1} of ash. While ash treatment has been the dominant waste processing application of these torches, they have also been used for treatment of sewage sludge, incinerator ashes, solid hospital waste and low level nuclear waste with a 480kW, 70 kg h^{-1} installation [31]. The twin-torch arrangement has the additional advantage of a more distributed heat load to the melt and, therefore, reduced evaporation of metallic components and formation of fine particulates.

Tetronics is also addressing the issue of energy reclamation from large waste streams. The Gas plasma process is combining fluidized bed gasification with plasma cleaning of the resulting hydrogen rich syn-gas. More than half of the heating value can be recovered in the form of syn-gas, and over 60% of the generated power can be exported. The plasma provides the high temperature environment for processing residual tars and chars, and allows vitrification of the ash into a non-leaching slag [61].

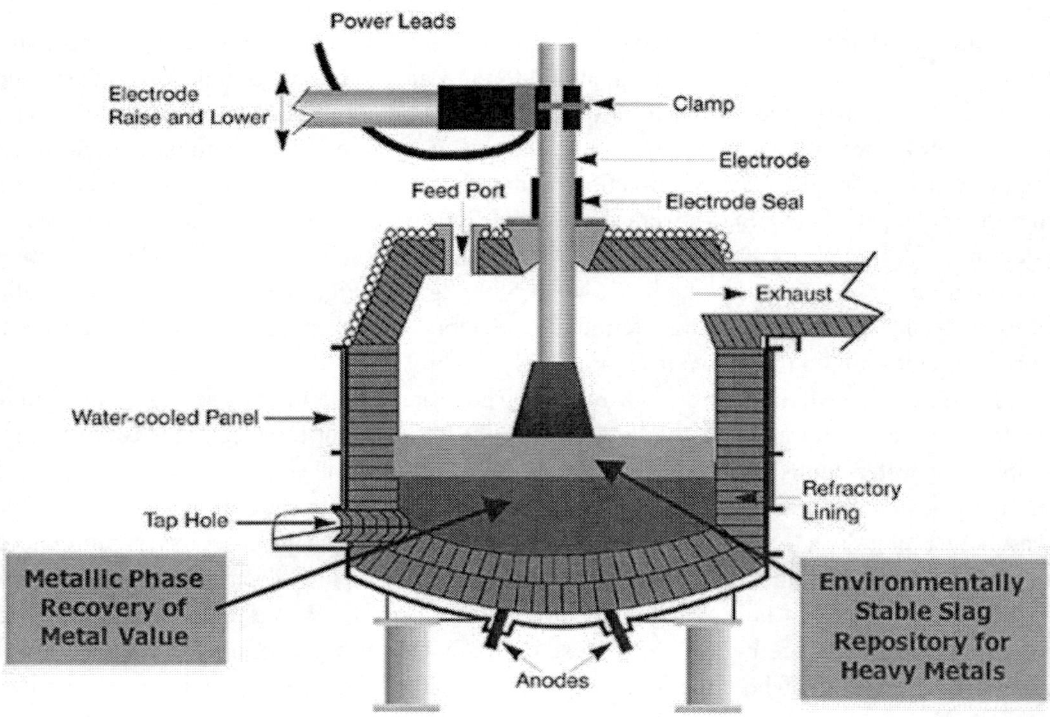

Figure 6. Tetronics single-torch transferred arc reactor.

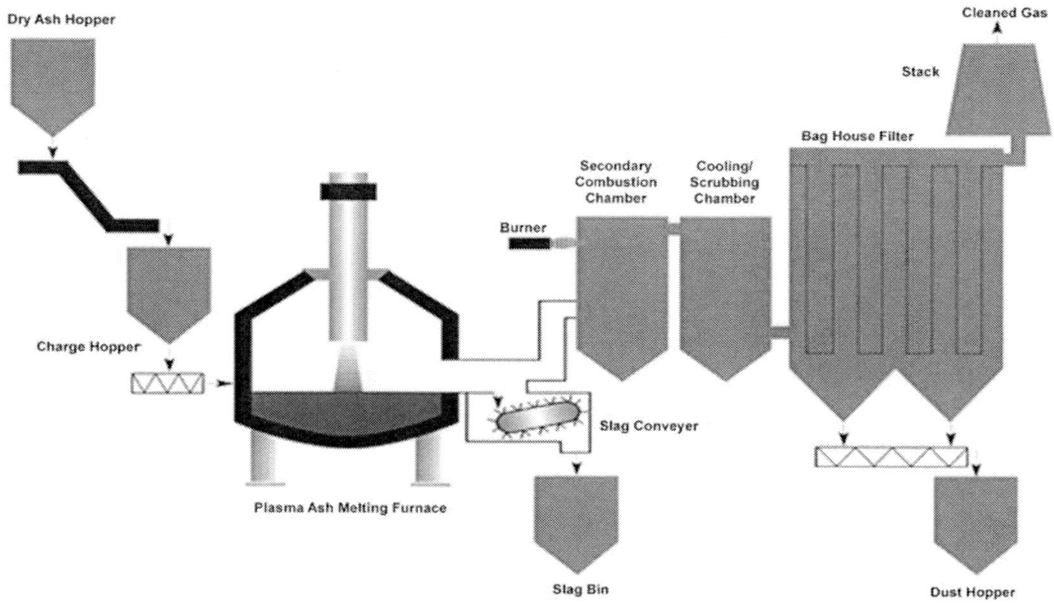

Figure 7. Process schematic for Tetronics ash vitrification process using single-torch transferred arc reactor.

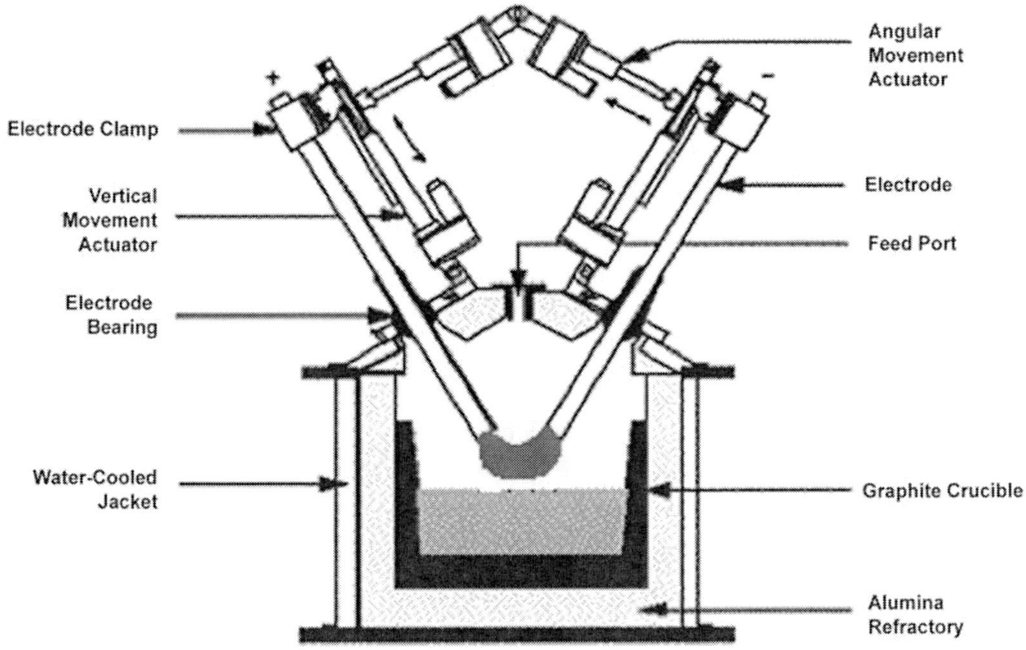

Figure 8. Schematic of Tetronics twin-torch reactor.

PHOENIX Solutions/PEC (Minneapolis, MN) has installed plasma systems for a number of ash vitrification plants, again the majority in Japan. Figure 9 [22] shows a typical flow diagram for such an ash treatment process. It should be mentioned that the torch is started in a non-transferred mode, and the arc is then transferred to the charge. These torches typically operate with the torch electrode serving as the anode and the charge as the cathode, and very long electrode life can be obtained this way. The power levels range from 150kW to 2MW, and five of the installations use multiple torches. Molten bath temperatures are given as >1500 °C, allowing addition of ceramic materials to allow casting of the slag into ceramic tiles with very low leach rates. This company has also an installation for asbestos treatment with a 500kW torch installed in a mobile reactor. A treatment rate of 250 kg h^{-1} has been achieved, with an immobilization of the asbestos fibres and a volume reduction by a factor of 12 [22].

Another interesting plasma system has been developed by RETECH Inc. (Ukiah, CA) based on a system used for metal remelting [62, 63]. While this system has been well described in the literature over the past 20 years, it was recently certified for treatment of low level nuclear waste with two commercial installations in operation. In this approach, the arc is transferred from a torch that can be operated also in the non-transferred mode, to the waste material which is contained in a rotating tub with an opening in its centre. At starting, the arc may operate in the non-transferred mode, or it may be transferred to a conducting refractory base in the centrifuge. The tub rotates at around 60 rpm. Periodically, the rotation is slowed and the liquid waste material is allowed to flow through the central hole into a crucible below. Figure 10 [64] shows the schematic of a waste processing installation.

The waste is contained unsorted in large drums and consists typically of a mixture of organic and inorganic waste material, and the drums are fed without being opened. The off-gas is treated first in a secondary combustion chamber to assure complete reaction of all organic material before going through a regular off-gas cleaning process. However, off-gas flow rates are very low because there is no need to stabilize the transferred arc. This system has been used by MGC Co. in Switzerland for the development of a low level nuclear waste compaction system, which is now operating in Zwilag (i.e. the Swiss nuclear waste depository) using a 1.2MW RETECH plasma reactor [65, 66]. A similar system is operated in Japan by the Japan Atomic Power Co.

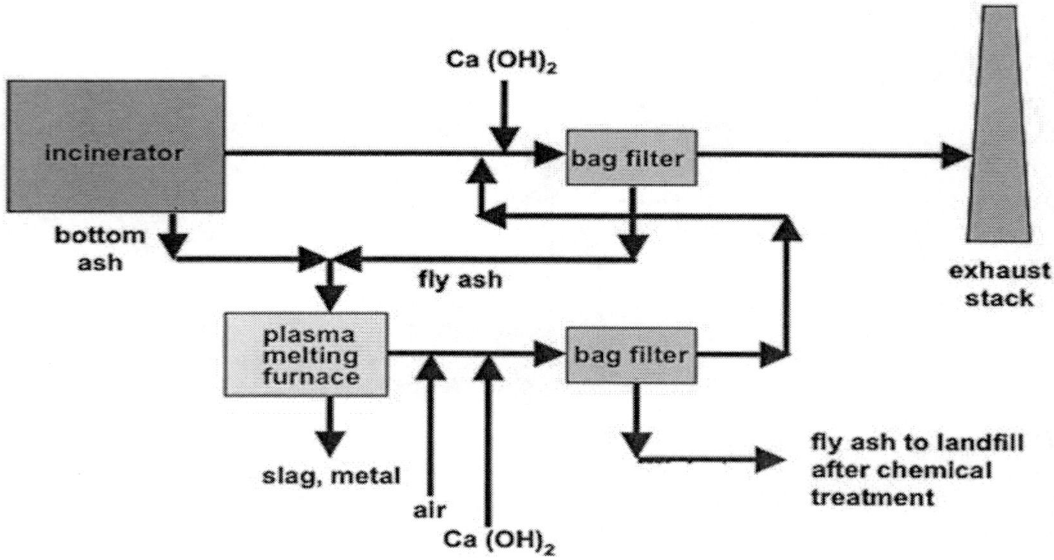

Figure 9. Plant lay-out for Phoenix Solutions ash vitrification installation.

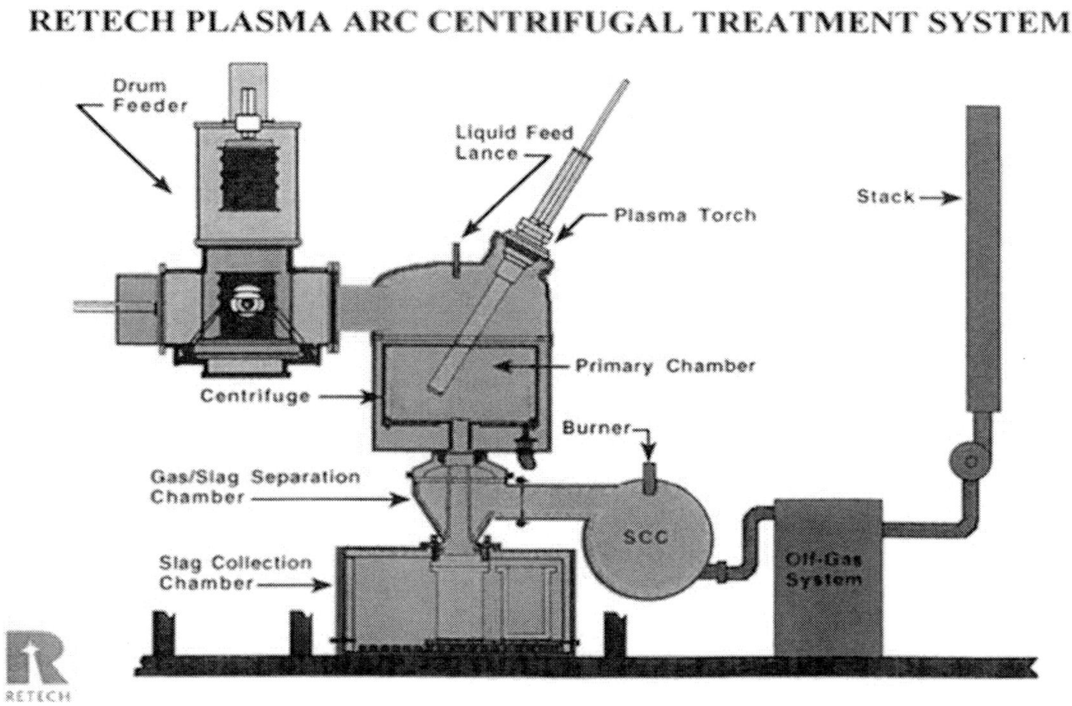

Figure 10. Schematic of the RETECH waste treatment installation using the PACT system.

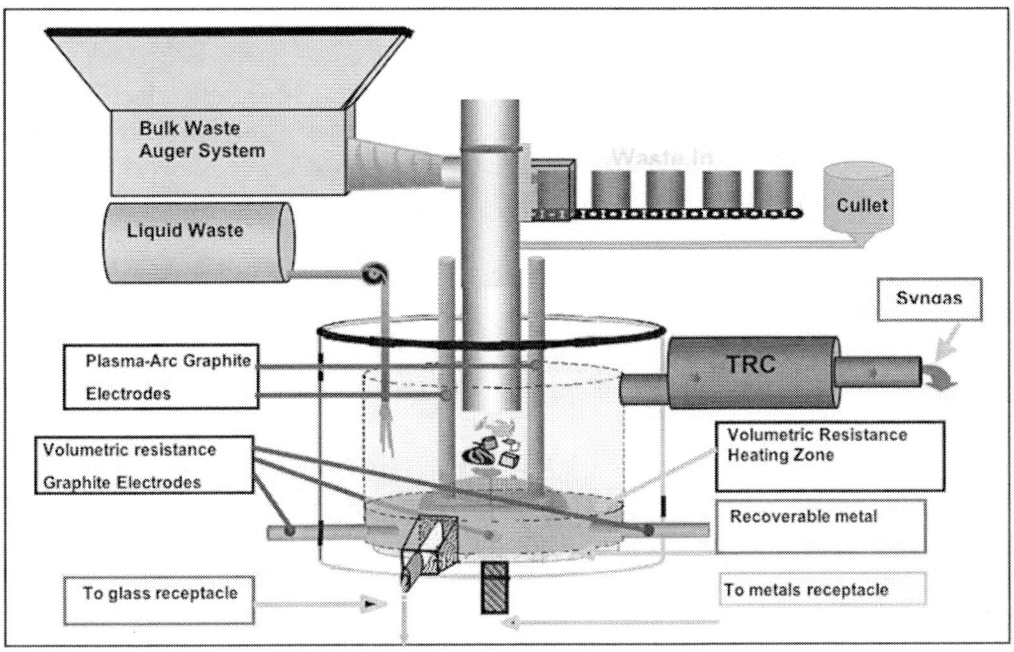

Figure 11. Schematic of the IET plasma enhanced melter (PEM) showing the two graphite arcing electrodes and the ac contacts for resistive heating of the molten material.

An interesting approach, the plasma enhanced melted (PEM), has been pursued by Integrated Environmental Technologies, LLC (IET, Richland, WA) [67]. The PEM process can treat a variety of wastes by a mixture of plasma heating and Joule heating. A dc arc is established between two graphite electrodes or between each electrode and the melt in a series configuration. Once the waste has become conductive, an ac current is passed through it heating and melting it resistively (see figure 11) [68]. The reactor is periodically tapped to remove the molten metal and the slag. The off-gas enters a thermal residence container to assure complete reaction of any volatile organic components, followed by a steam quench and a filtration and scrubbing system.

The syn-gas produced consists typically of 30–50% hydrogen and 30–40% of CO, depending on the type of waste, and can be used for generation of pure hydrogen. The low level of nitrogen (around 6%) is due to the fact that the atmosphere in the plasma chamber consists of oxygen rather than air. Two basic systems are offered, with nominal processing rates of 10 t d^{-1} and 17 t d^{-1}, respectively. The ac power used for Joule heating is 150kW and 225kW, respectively, for the two systems, and comparable power values have to be added for utilities (e.g. pumps, etc). The dc power for the plasma torches is typically between 150 and 900kW, and depends on the plasma gas and on the type of waste. Correspondingly, the specific power requirement is 1100 kWh t^{-1} for medical waste (this includes the utility power; only 300 kWh t^{-1} is required for the plasma power). Typical energy content of the syn-gas is of the order of 10 MJm^{-3}. It is estimated that for waste throughputs of 6–10 t d^{-1} enough electricity can be generated to operate the installation. Several commercial systems treating up to 17 t d^{-1} are in operation, including one in Richland, WA, for treating LLRW [68].

Pyrogenesis has also developed a process which combines vitrification of inorganics and gasification of the organic waste material and has tested it in pilot plant operation with a variety of waste materials [32, 49, 50, 69]. In this plasma resource recovery system, the shredded waste material is introduced into an arc furnace with two graphite electrodes. The plasma volatilizes the organic material and vitrifies the inorganic material. The off-gases are mixed with air plasma and steam in a plasma fired eductor, as used in the ship board waste treatment. This way, unreacted organic material as well as soot is gasified. Particulate filters and quenching complete the installation. The resulting gas composition for an MSW treatment test run consisted of about 20–29% CO and 8–14% H_2, with most of the remaining gas being nitrogen. This gas has been used for electric power generation in gas engines. Performance data projected from pilot plant operation for a 240 t d^{-1} MSW plant are 843 kWh t^{-1} specific energy requirement; 3900kW is required to operate the reactor with all the ancillary equipment, resulting in a net power production of 4530kW [32].

5. Hydrogen Recovery from the Thermal Plasma Gasification of Solid Waste

The recovery of energy from the increasing amounts of solid waste would be very useful as a supply of sustainable energy. Therefore, a more recent trend for the treatment of solid waste is the combination of incineration and energy recovery in the so-called "waste-to-energy" plants [70]. This combination helps to solve two problems, the first is to provide the energy, and the second consist of solving the environment problems. Therefore, thermal gasification of solid waste has been proposed and developed as an alternative solution for the destruction of waste with the simultaneous recovery of energy [71, 72].

Plasma gasification technology has the following advantages: (1) the high energy density and temperature, (2) the treatment of a wide range of waste, and (3) the high heat flux density at the reactor boundary that lead to the fast attainment of steady state condition [73, 74]. With the plasma gasification, the organic components of waste are converted to syn-gas, mainly composed of carbon monoxide (CO) and hydrogen (H_2), which can be used directly as a source of fuel in power plants for the production of electricity or synthetic fuels [75]. Therefore, over the past decade, the thermal plasma process has been regarded as a viable alternative for the treatment of highly toxic waste, such as the residue from MSW incineration (bottom ash + fly ash), radioactive, and medical wastes [76- 84].

Methods for the recovery of energy from the gasification of solid waste include: (1) manufacturing fuels or chemicals from the syn-gas, (2) the recovery and utilization of H_2 as a fuel, (3) the recovery of H_2 and the production of electricity using fuel cells, and (4) the production of electricity using a complex gas-steam turbine cycle [85]. Of these, it is worth noting that the recovery of H_2 from the gasification process has increased, which has motivated the development of H_2 recovery technology in several investigations using the thermal plasma treatment of solid waste, while no purification process has been conducted for the recovery of high purity H_2 [86, 87].

In the following paragraphs of this section we present the materials, methods, and results, obtained by Byun Y., et al., during their study on hydrogen recovery from thermal plasma gasification of solid waste [88].

5.1. Materials and Methods

The schematic diagram and a picture of the thermal plasma gasification/H_2 recovery systems are shown in figure 12. The process was composed of two sub systems: a thermal plasma gasification system, which converts the organic components of the feed waste into syn-gas, and an H_2 recovery system, which converts the syn-gas generated by the thermal plasma gasification of waste into high purity H_2 (>99.99%) using WGS and PSA.

Thermal Plasma Gasification System

The thermal plasma gasification system was a 3 tonnes/day scale apparatus (maximum capacity), consisting of a waste feeding system, a thermal plasma furnace equipped with a non-transferred plasma torch, a heat exchanger, flue gas treatment units, and a post-combustion chamber. A pressurized ram type feeder was used to supply waste into the plasma furnace. Approximately 1–2 kg of waste, enveloped in a plastic bag, was placed in the entrance slot, and then pushed into the plasma furnace continuously by a 10 cm diameter piston. The waste feed rate was about 50 kg/h.

A non-transferred plasma torch, with a power capacity of 100 kW, was installed in the plasma furnace at an angle of 20°. The torch angle induced cyclonic flows of the gas streams to allow effective gas mixing and prevent the emission of fly ash [74]. The operational voltage and current of the plasma torch were maintained to 350±10V and 200 A, respectively; the energy efficiency of the plasma torch was 70%. O_2 (12Nm^3/h at 506.5 kPa), which was produced by a two stage PSA reactor packed with a zeolite molecular sieve and activated carbon on sites with a purity of 93%, was injected into the plasma torch. An auxiliary gas burner, which used liquefied natural gas (LNG), was also installed at the top of plasma torch to preheat the furnace (LNG: 2Nm^3/h and air: 20Nm^3/h).

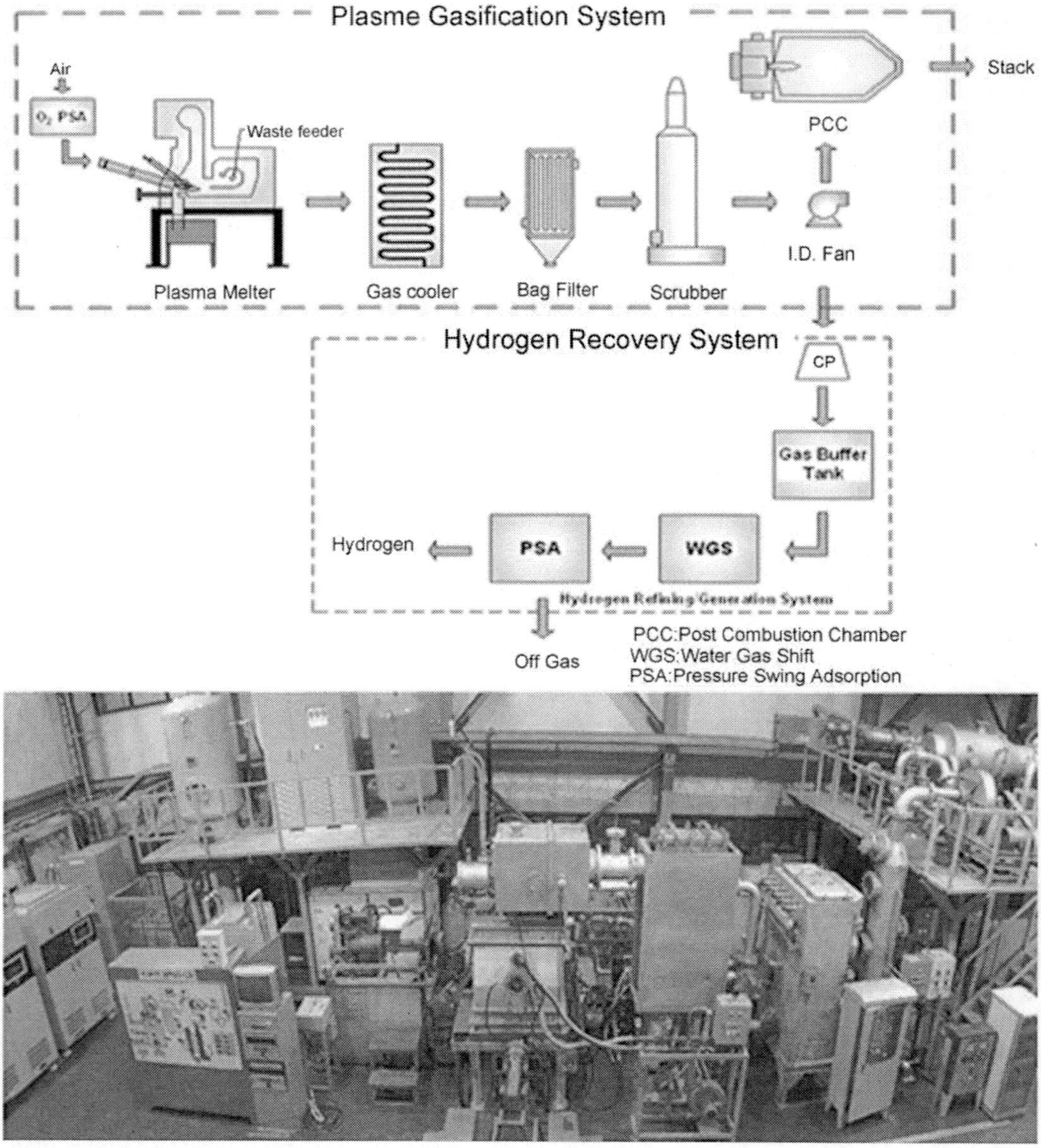

Figure 12. Schematic diagram of the thermal plasma process for the recovery of high purity H_2, and the bottom picture has shown the overall demonstrated plant.

The temperature in the plasma gasification furnace was maintained at 1400–1450 ˚C. An outlet for the molten slag was placed at the bottom of the furnace. The heat exchanger played the role of decreasing the flue gas temperature to about 180 ˚C. A bag filter and wet scrubber were used to eliminate particulate matters and acidic gases produced from the thermal plasma gasification. The bag filter was composed of 18 filters and a gas pulsing system to remove the particles adhered to the bag surface. An inert gas, either N_2 or CO_2, was used as the bag filter pulsing gas to prevent the syn-gas from exploding. Caustic soda (NaOH) was used to adjust the acidity of the scrubbing solution to pH 7. The syn-gas produced was supplied to the H_2 recovery system or post-combustion chamber which was equipped with a gas burner to combust the syn-gas.

H_2 Recovery System

The H_2 recovery system was mainly composed of the WGS and the PSA connected in series. The gases entering the WGS were compressed to nearly 700 kPa. Two stage WGS reactors were used, connected in series: one was a high temperature shift (HTS) reactor operated at about 350 °C, with the other being a low temperature shift (LTS) reactor operated at approximately 200 °C. The HTS and LTS reactors were filled with Fe/Cr type and Cu/Zn type catalysts, respectively. In addition, a guard bed for the removal of H_2S was installed between the HTS and LTS as the catalyst packed in LTS can be deactivated by the deposition of sulfur [89].

The role of the PSA system was to produce high purity H_2 by separating the impurities by the different volatilities and polarities of the gas components. The PSA system was composed of four-bed reactors operated in parallel. The adsorption and regeneration cycles were continuously repeated in a controlled sequence to generate high purity H_2. The H_2 produced was also used as a purging gas to regenerate the PSA adsorbent after the adsorption cycle. The operating pressure of the PSA reactor was controlled by a gas compressor installed in front of the H_2 recovery system. The optimization of H_2 recovery system was conducted focusing on the PSA unit, since the performance of the WGS system was stable throughout the entire experimental procedures.

Waste Characteristics and Gas Analysis

The waste generated from the process of raw materials in a paper mill plant was used as the feed material for the recovery of H_2; the chemical characteristics of this material were similar to those of biomass reported in the literature [90– 93]. The waste was made up of 68.8% combustibles, 20.6% moisture, and 10.6% non-combustibles. The most abundant element was carbon (46.8%), followed by oxygen (33.7%), hydrogen (7.5%), nitrogen (0.9%), chlorine (0.1%), and sulfur (0.02%). The lowest heating value of the waste was 3300 kcal/kg. Almost all the operational processes, such as the operational voltage, O_2 supply, cooling water supply for the plasma torch, the air pollution control system, and H_2 recovery system, were automatically controlled and monitored. A gas analyzer and gas chromatography were used to analyze the gas components.

5.2. Results and Discussion

Five sets of experiment were carried out over a 6 month period using paper mill waste PMW, focusing on the compositions and environmental safety of the gases generated from the thermal plasma gasification system and the optimization of the H_2 recovery system (WGS+ PSA units) for the production of high purity H_2 (>99.99%).

The concentrations of CO_2, CO, and H_2 were measured at the wet scrubber outlet; the flow rate of the gases was 65.2Nm^3/h during the five experimental sets. The total average concentrations of CO and H_2 were 39.56% and 34.90%, respectively, resulting in a sum

74.46% for the two components (table 1). These results suggest the production rates of CO and H_2 from the inlet PMW were 516 and 455Nm3/PMW-ton, respectively. In addition, the concentration of CO_2 emitted during the five experiments was 12.6% (164Nm^3/PMW-ton). From the concentrations of CO and CO_2, we calculated the carbon conversion efficiency as follow;

$$carbon\,conversion\,efficiency(\%) = \frac{total\,reacted\,carbon\,in\,the\,system\,(kg)}{total\,carbon\,fed\,in\,the\,system\,(kg)} \times 100$$

The total amounts of reacted carbon in the system were obtained from the CO and CO_2 concentrations generated. As a result, a 97% carbon conversion efficiency was obtained. This result suggests that the process of thermal plasma gasification completely gasified the solid waste.

Table 1. Concentration of emitted CO and H_2 measured at the outlet of wet scrubber according to the experimental set [69]

	Avg. CO (%)	Avg. H_2 (%)	Sum of CO+H_2 (%)
Exp. 1	38.77	35.21	73.98
Exp. 2	40.58	34.65	75.23
Exp. 3	37.65	33.93	71.58
Exp. 4	40.38	35.60	75.98
Exp. 5	40.44	35.09	75.53
Total avg. (%)	39.56	34.90	74.46

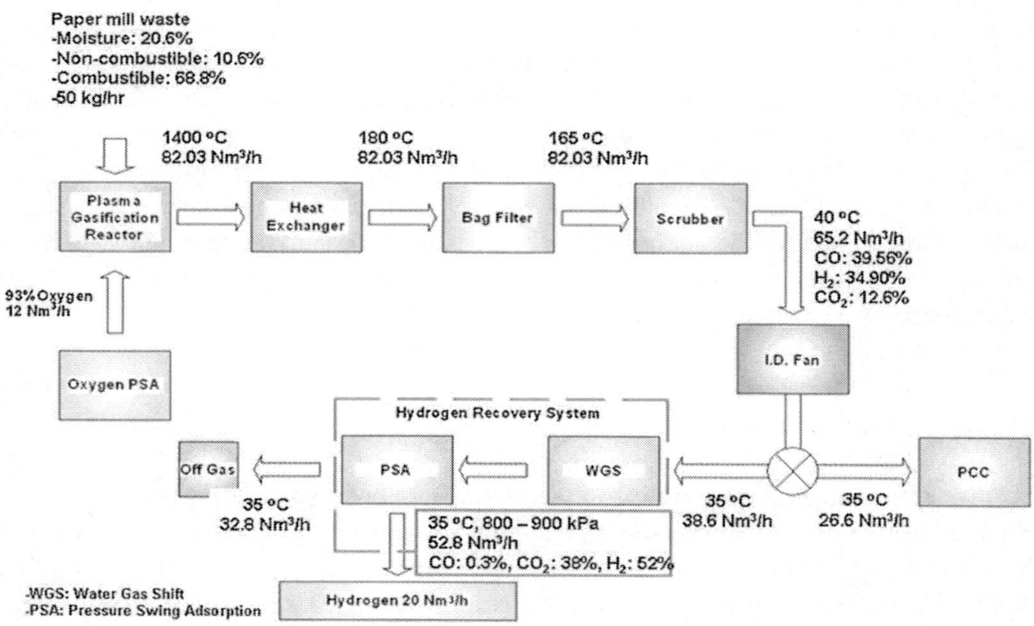

Figure 13. Material balance of the overall process for the recovery of high purity H_2 using thermal plasma.

6. Long- term Behavior of Vitreous or Crystallized Materials Obtained by Vitrification of Solid Wastes

Long-term behavior studies with vitreous or crystallized materials obtained by vitrification of domestic waste incineration fly-ash purification residues have demonstrated a high degree of containment of toxic elements within the vitrified product [94]. Rather than constituting simple landfill, these materials may therefore be suitable for re-use. From this perspective, however, it is indispensable to provide the assurance of their resistance to alteration under any circumstances. In this context, different classes of vitrified product have been defined. A vitrified product class is "a group of vitrified products belonging to a given composition range and fabricated under specified conditions, resulting in the same type of long-term behavior in a given disposal scenario" [95]. As the fly-ash residue is generally vitrified without additives, any composition variations will directly affect the composition of the final material.

In this section we will detail the method and the results obtained by Frugier et al. [96 - 97] during their study on the Influence of composition variations on the initial alteration rate of vitrified domestic waste incineration fly-ash.

6.1. Composition Variations

In addition to random or periodic variations, and systematic variations specific to a given incineration plant, two types of fly-ash residue must be distinguished: ashes resulting from dust scrubbing after simple filtration, and ashes containing larger quantities of lime produced by combined treatment of acid off-gases by a wet or dry process [98]. The high temperatures (>1400 °C) necessary to melt fly-ash residues and their chemical composition (chlorides, reducing atmosphere over the melt) result not only in virtually complete volatilization of the chlorine and sulfur, but also in appreciable volatility of some other elements, mainly the alkali metals and heavy metals. Without effective recycling of the cations in the second generation ashes recovered in the process off-gas treatment system, the vitrified product composition is modified accordingly.

6.2. Vitrification Tests

The first step was to ensure the quality of the vitrified product, i.e. a solid obtained by cooling a super cooled liquid and liable to contain a variable fraction of phases crystallized during cooling; glass is a vitrified product. Unsatisfactory materials can be identified by an excessive soluble fraction resulting in unmelted inclusions. Ten melting tests were carried out in the laboratory to characterize a wide composition range for a preliminary performance assessment, revealing several important points [99]:

- A high calcium content in the ashes results in a correspondingly lower fraction of the essential glass network-forming element silicon, and reduces the likelihood of obtaining a suitable vitrified product.

- For the same reasons, high alkali metal concentrations in this composition range—which is already rich in network modifiers—prevents vitrification. The plasma arc melting process developed by the French electrical utility EDF [100] considerably increases the chances of obtaining suitable vitrified products by raising the volatility of the alkali metals to over 70% [99].
- The tests showed that vitrified products are obtained by melting any combination of oxide powders from the composition range of fly-ash residues obtained by simple dust scrubbing, provided the process induces at least 50% alkali volatility.
- The crystallization kinetics, notably for gehlenite and chromiferous spinels, were highly variable over the studied composition range. Their occurrence and their effect on leaching behavior were examined in another study [99].

The influence of composition variations on the alteration kinetics can be investigated only within a composition range in which the material structure remains similar. This study is therefore limited to vitreous materials, i.e. all the vitrified products in the composition range that produce glass when cooled at a sufficiently high rate.

6.3. Choice of Composition Parameters

The elements assumed to have the greatest effect, considering their respective variation ranges, were the following:

- the three major elements: silicon, calcium and aluminum;
- two alkali metals: sodium and potassium, with wide variation ranges due to the effects of the process, and with potentially major effects on the characteristic glass parameters; and
- two toxic elements: zinc and lead, selected in view of the effects of the process operating conditions, which are more significant, even than for the alkali metals. Only lead and zinc are found in the composition in sufficient amounts to affect the behavior of the matrix; they were selected as well in response to concerns by industrial operators on the role of these elements in processes operating with or without recycling of second-generation ashes.

Fly-ash analyses revealed a strong correlation between the alkali metals and zinc, lead or even cadmium; these elements were therefore included in the study. This simplification is further justified by the comparable chemical behavior of the alkali metals in the glass. Seven elements (Si, Al, Ca, Na, K, Pb and Zn) were therefore selected as parameters, together with two constraints: the molar ratios of Na/K (2.0) and Zn/Pb (8.2).

The elements maintained at constant (mean) concentrations accounted for less than 10 mol% of the total cation content: Mg, Fe, Ti, P and Cr (Table 2). The numerous minor elements inevitably present in this type of waste represented less than 0.2 mol% and were therefore not taken into account. The selected chromium content was higher than found in actual fly-ash residue; this value was used to allow for possible salting-out of chromium from the refractory furnace walls into the glass melt during the process. Lithium was added as a corrosion tracer. As the added lithium contributed to the total alkali metal content, the maximum concentrations proposed for Na and K in Table 2 were reduced accordingly, while maintaining the Na/K stoichiometry (the important parameter is the total alkali metal content).

In his study Frugier [99] retained two scenarios: the first called "pure water leaching" it can measure the speed of alteration for a given temperature at neutral pH. Whereas the second scenario considers the leaching of closed system at 25 ° C and pH = 9 ; which is closer to the storage reality and it has the advantage of exceed the dissolution rate for any other scenario where the temperature would be below 25 ° C and pH between 6 and 9. Frugier has proposed four classes of vitrified presented in figure 14. Classes were noted from I to IV according to the release amplitude that they induce. The domain 1 defines Class I, domain 2a class II (glasses) and III (slag crystallized) and the domain 2b class IV.

Table 2. Experimental composition range (cation mol %)

Elements	Si	Al	Ca	Na	K	Pb	Zn	Mg	Fe	Ti	P	Cr	Li
Lower limit	26.0	15.1	25.8	0	0	0	0	4.4	1.5	1.4	1.3	0.4	3.6
Upper limit	42.5	30.8	34.5	4.6	2.3	0.4	2.8	4.4	1.5	1.4	1.3	0.4	3.6

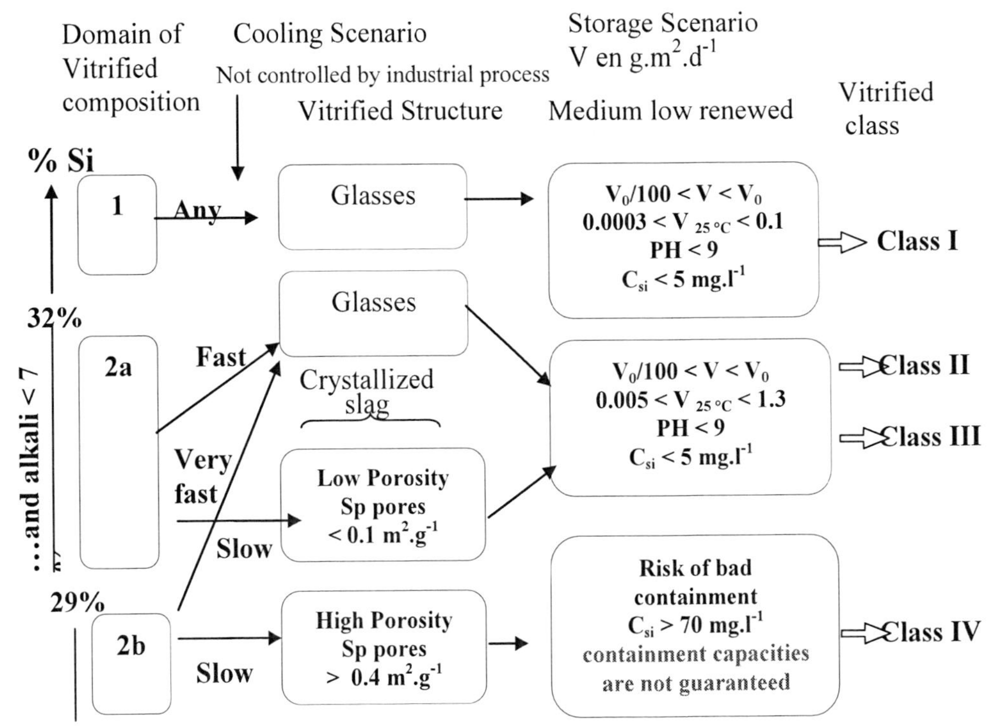

Figure 14. Classes of vitrified in a leaching of closed system.

The initial alteration rates of vitrified fly-ash residue in pure water at 100 °C varied by an order of magnitude, between 4 and 34 $g.m^{-2}.day^{-1}$. The rates increased as the alkali content increased and the silicon content decreased in the material. The toxic elements lead and zinc had virtually no effect on the initial alteration rate of the vitrified product. The initial alteration rate in pure water at 100 °C of a glass with a composition from within the studied range can be guaranteed to be less than 40 g m^{-2} day^{-1}. It is important to note, however, that the initial rate alone is not sufficient to define classes of vitrified products, or even to predict subsequent material behavior under the complex conditions prevailing in a repository environment.

From the study of Frugier [99], only the material containment capacity of class IV has insufficient performance. Indeed, this slag may have alterability very different and potentially very high for small variations in composition. In addition, the range of composition of this slag is very limited and far from the compositions of industrial slag, which is depleted in alkaline. On Classes I, II and III, the achieved containment is very good.

Two domains of composition are distinguished according to the influence of crystallization on the long-term behavior:

- For molar Silicon contents greater than 29 % and molar alkali content less than 7 %, the variations in speed caused by the crystallization are of the same order of magnitude as those due to changes in composition of glasses. In first approximation, we can assimilate the behavior of these glasses to that of crystallized slag having same composition. The crystals formed are more resistant to alteration than the glass. Spinels, in particular, favor the containment of zinc and chromium that constitute them.
- For molar Silicon contents less than 29 % (or molar alkali content greater than 7 %) the crystallization is immediate: it appears even for a very fast cooling scenario. it leads to a porous material whose contact surface with the solution is considerably increased. The coupled effects of porosity and composition of the residual glass phase (more soluble as alkali-enriched and depleted in silicon) significantly lower containment capacity of these materials.

Generally, slag crystallized from industrial processes realizing a high-temperature melting and studied to date, are sufficiently depleted in alkali (volatile) to belong to the first composition domain described above.

7. Ewaste Treatment by Plasma

The investigations on electronic waste treatment with plasma are still very rare. Among these few studies, we can mention that made by Byun [101], which we will present in this section. Most studies focused on using inert gases or steam as working gasses. Due to economic reason, use of the cheapest gas (the air) appears most practical. Therefore, this investigation is an attempt to lessen this gap [101].

In this study, a 20 kW atmospheric air plasma using DC discharge was developed for a laboratory study. The objective was to investigate DC plasma characteristics and utilize it to process electronic waste. Figure 15 shows schematically design of a DC plasma torch, consisting of a steel rod cathode and an air cooled anode, shaped in the form of nozzle. The two electrodes are separated by an insulator which has an inlet for plasma gas. When compressed air is introduced thru the electrode gap and a DC arc is established between the electrodes, the plasma arc is emerged from the nozzle resulting in a high temperature, high velocity plasma flame.

Electromagnetic forces from magnetic coil and gas flow act to constrict the arc column. The body of the torch was designed such that there are cooling chambers for cathode and anode. The torch was coupled to the main power supply and operated in a non-transferred arc mode. Air or a mixture of air and other gases can be used as medium gas. Liquefied petroleum gas (LPG) can be injected to air plasma via connecting pipe. For the preliminary design, the operation ranges of power and air flow of the plasma torch were limited to 12-20 kW and 300-l pm, respectively. Temperatures were measured at different positions of the flame by a type K thermocouple. These locations were linearly 10 or 20 mm apart in axial direction and 5 or 10 mm apart in radial direction of the main flame column, starting from the torch exit.

A schematic diagram of a laboratory scale, plasma furnace for treatment of electronic waste is shown in figure 16. It was equipped with the air plasma torch, a refractory-lined circular reactor and a gas flare prior to exhaust. The dimensions of the reactor were about 100 mm in diameter and 200 mm long. Old central processing unit (CPU) board was used as representative electronic waste. The waste of 0.2 kg can be batch fed into the reactor where rate of reduction in mass and the reactor temperature were monitored.

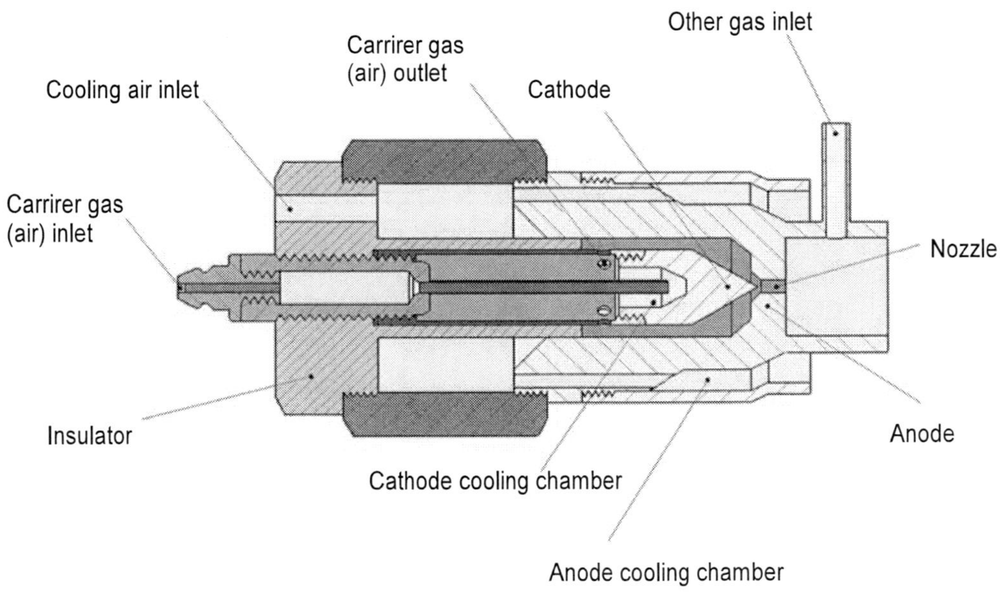

Figure 15. Schematic drawing of a DC plasma torch.

In this experiment, CPU board of 0.2 kg was thermally decomposed in the reactor. Measurement of temperature evolution inside the furnace was performed. It should be noted that the plasma torch was fully emerged inside the furnace and the temperature probe was positioned at the reactor axis, 100 mm away from the nozzle. It was evident that the reactor temperature increased at a rapid rate and leveled off around 1,200 and 1,100 K at 50 and 100 mm, respectively. The temperatures at both locations were close but did not attain the maximum level reported previously in the naked plasma flame. The Electronic waste can be processed using high temperature air plasma torch operated at 20 kW rating.

It can be seen that more than 50 % of solid electronic waste was reduced in less than 120 s and the treatment was completely finished in about 150 s. Most were rapidly converted to gases, while about 20 % of the initial mass remained as solid residues. Change in physical appearance of the CPU board sample with time was illustrated in figure 17. It was clear that the air plasma can disintegrate, completely destroy and reduce volume of solid electronic waste. The remaining solid residues were inert materials that can be easily disposed of, or used as part of bulk material in road construction.

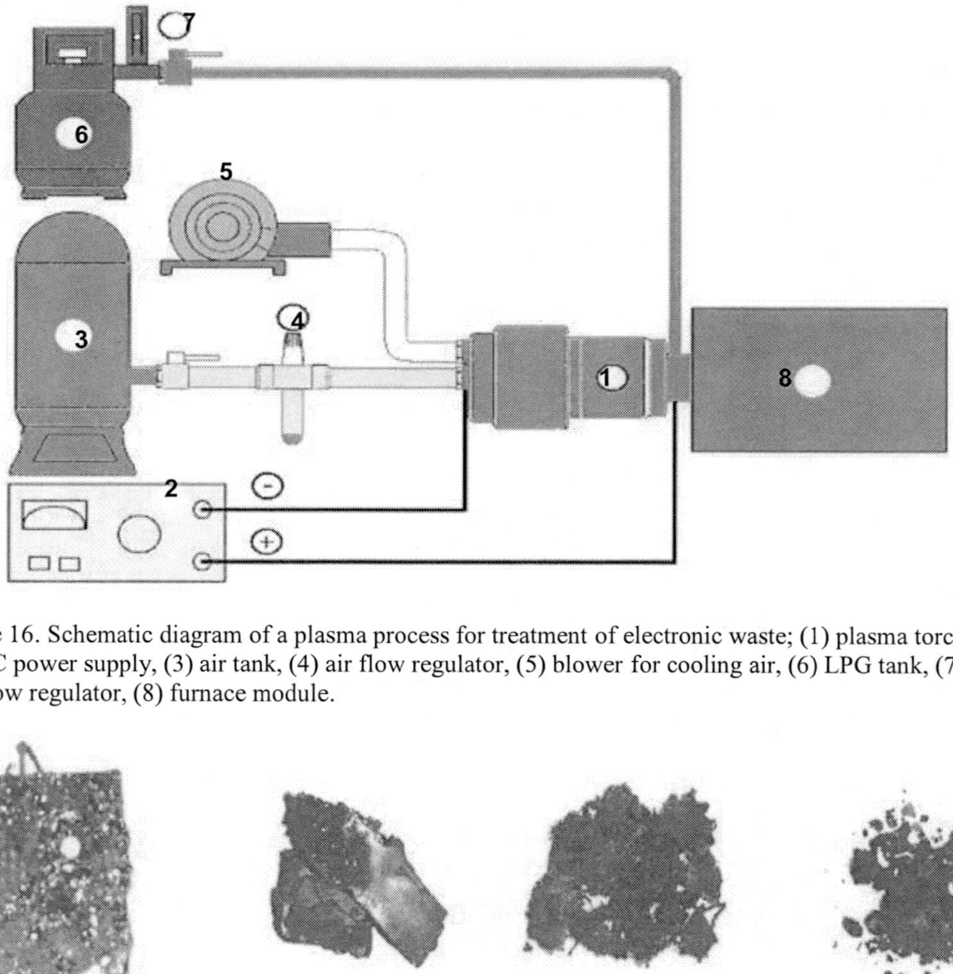

Figure 16. Schematic diagram of a plasma process for treatment of electronic waste; (1) plasma torch, (2) DC power supply, (3) air tank, (4) air flow regulator, (5) blower for cooling air, (6) LPG tank, (7) gas flow regulator, (8) furnace module.

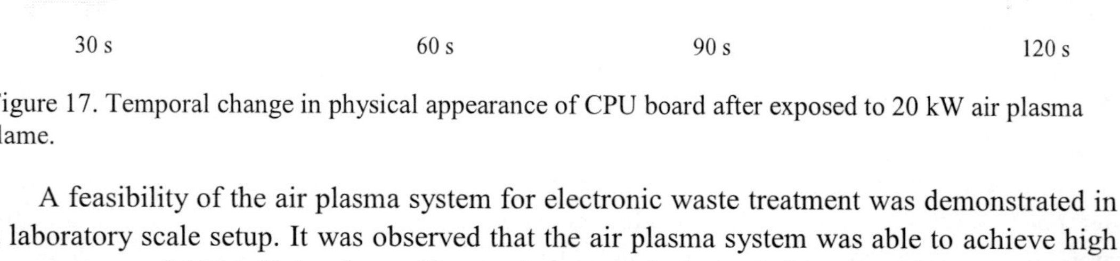

30 s 60 s 90 s 120 s

Figure 17. Temporal change in physical appearance of CPU board after exposed to 20 kW air plasma flame.

A feasibility of the air plasma system for electronic waste treatment was demonstrated in a laboratory scale setup. It was observed that the air plasma system was able to achieve high temperature of 1200 K in about 80 s and promoted thermal decomposition and volume reduction of the waste to below 80 % in just over two minutes. A number of trial runs demonstrated that the non-transferred DC air plasma was effective for the destruction of electronic wastes. Technically, plasma technology can be adopted as an alternative electronic waste treatment for fast operation and elimination of slag toxicity. However, electrical energy and cost of electricity involved are still the main obstacles towards utilization of the plasma waste treatment technology. The investigation of ways to improve the plasma torch performance and increase the efficiency of the process is very crucial.

CONCLUSION

Expansion in electrical and electronic equipment trade has led to significant increase in electronic waste which should be dealt with special priority due to its potential negative impact to the public health and the environment. Treatment of e-wastes has become one of the central problems of our age, as traditional e-waste repositories, such as landfills and ocean dumpsites, become less and less available. Many countries have introduced tight regulations that restrict dumping in landfills. Since many types of e-wastes are combustible, incineration can be used for reduction of these wastes; however, high off-gas flow rates and the consequent high cost of off-gas clean-up, as well as the production of residues containing hazardous materials, mean that incineration does not offer a complete answer to the e-waste problem. Furthermore, certain components of some electronic products contain materials that render them hazardous, and some compounds such as chloride and heavy metals can remain in fly-ash and cause secondary environmental problems. For this reason, alternative thermal treatment technology such as thermal plasma is required for sterilization.

Compared with other waste treatment technologies, the principal advantages of plasma technology are the reduction of exhaust gas flow rates, in most cases higher volume reduction, installations with smaller footprints and lower investment costs for a given throughput, and faster start-up and shut-down times. Clearly, plasma treatment is advantageous where landfill is difficult or impossible, where storage costs are encountered, or where stringent emission standards make combustion-based treatment expensive. This is usually the case for hazardous materials, in particular the fly ash resulting from the incineration of e-wastes.

Plasma e-waste treatment can be a more prominent technology because of the increasing problems with e-waste disposal and because of the realization of opportunities to generate valuable co-products. For the waste majority, economic considerations dictate the combined use of several treatment technologies. The realization that waste streams can be a source of energy, and that plasmas can be useful to tap this source through controlled gasification processes can increase the number of plasma process developments. Here, the use of plasmas is attractive when a valuable co-product can offset the negative value of the e-waste augmented by the treatment cost. Generations of syn-gas, hydrogen or electricity are prime examples of such co-products.

Long-term behavior studies with vitreous or crystallized materials obtained by vitrification of domestic waste incineration fly-ash purification residues have demonstrated a high degree of containment of toxic elements within the vitrified product. Rather than constituting simple landfill, these materials may therefore be suitable for re-use. The disadvantages associated with the plasma process, besides the already mentioned (largely perceived) lack of reliability, lie in the use of electricity, the most expensive form of energy. Consequently, economic considerations provide the strongest barrier for use of plasmas for e-waste treatment. For this reason, plasma gasification of e-wastes with low negative values can be interesting as a source of energy and spawned process developments for e-waste treatment.

REFERENCES

[1] Hsu C. W., Hu A. H., (2008). *Int. J. Environ. Sci. Tech.,* 5 (2), 205-216.

[2] Nnorom I. C., Osibanjo O., (2008). Overview of electronic waste (e-waste) management practices and legislations, and their poor applications in the developing countries. *Resour. Conser. Recycl.,* 52 (6), 843-858.

[3] Widmer R., Oswald-Krapf H., Sinha-Khetriwal, D., Schnellmann M., Boni H., (2005). Global perspectives on e-waste. *Environ. Impact Assess. Rev.,* 25 (5), 436- 458.

[4] Shekdar A. V., (2009). Sustainable solid waste management: An integrated approach for Asian countries. *Waste Manage,* 29 (4), 1438-1448.

[5] Bonizzoni G., Vassallo E., (2002). Plasma physics and technology; industrial applications. *Vacuum,* 64 (3-4), 327-336.

[6] Chang J. S., (2001). Recent development of plasma pollution control technology: a critical review. *Sci. Tech. Adv. Mater.,* 2 (3-4), 571-576.

[7] Kogelschatz U., (2004). Atmospheric pressure plasma technology. *Plasma Phys. Contr. F.,* 46 (12B), B63-B75.

[8] Leal-Quiros E., (2004). Plasma processing of municipal solid waste. *Braz. J. Phys.,* 34 (4B), 1587-1593.

[9] Pfender E., (1999). Thermal plasma technology: *Plasma Chem. Plasma Proc.,* 19 (1), 1-31.

[10] Ganguli A.; Tarey R. D., (2002). Understanding plasma sources. *Curr. Sci.,* 83 (3), 279-290.

[11] Venkatramani N., (2002). Industrial plasma torches and applications. *Curr. Sci.,* 83 (3), 254-262.

[12] Heberlein J., (1992). Thermal plasmas for the destruction of hazardous wastes *Proc. Plasma Technologies for Hazardous Waste Destruction (Como, Italy)* ed G Bonizzoni *et al* (Bologna, Italy: Editrice Compositori) pp 59–76.

[13] E. Gomez, D. Amutha Rani, C.R. Cheeseman, D. Deegan, M. Wise, A.R. Boccaccini, Thermal plasma technology for the treatment of wastes: A critical review, *Journal of Hazardous Materials* 161 (2009) 614–626

[14] Bruce K.R., Lee J., Freed D., Heredy L., (1998), "Evaluation of plasma – based thermal treatment system for destruction of difficult to remediate waste", *International Conference on Incineration and Thermal Treatment Technologies*, Salt Lake City, USA, pp.559-564.

[15] Springer M.D., Barkley T., Castellon O., Forsberg B., Stutts G., (1995),"A thermal destruction and recovery process for medical waste powered by a thermal plasma arc torch", *International Symposium on Environmental Technologies,* Atlanta, USA, pp. 671-682.

[16] Guihard B., (2002), "Industrial and environmental applications of non transferred plasma", *International Conference on Incineration and Thermal Treatment Technologies,* New Orleans, USA.

[17] Circeo L.J., Camacho S.L., Jacobs G.K., Tixier J.S., (1994), "PRISM – Plasma Remediation of In Situ Materials and its potential for the remediation of Chernobyl consequences", *33th Hanford Symposium on Health and The Environmental Symposium on in-situ remediation,* Richland, USA, pp. 1124-1128.

[18] J. Heberlein and A. Murphy, (2008), Thermal plasma waste treatment, *J. Phys. D: Appl. Phys.* 41 053001.

[19] I. Lopez, L. Borzacconi, Final disposition of municipal solid waste: need for an integral approach, *Ingenieria Quimica* 18 (2000) 45–52.

[20] J. Scott, D. Beydoun, R. Amal, G. Low, J. Cattle, Landfill management, leachate generation, and leach testing of solidwastes in Australia and overseas, *Crit. Rev. Environ. Sci. Technol.* 35 (2005) 239–332.

[21] L. Heller, C.A.A. Catapreta, Solid waste disposal in urban areas and health—the case of Belo Horizonte, Brazil, *Waste Manage. Res.* 21 (2003) 549–556.

[22] Hanus G 2007 Phoenix Solutions Co., personal communication

[23] Chapman C D, Williams J K, Heanley C P, Iddles D M, Forde A J and Risdon G 1995 Development of a plasma reactor for the treatment of incinerator ashes *Proc. Int. Symp. on Environmental Technologies: Plasma Systems and Applications (Georgia Tech Research Corp; Atlanta, GA)* pp 129–40.

[24] Heanley C P, Chapman C D and Wolf C J 1998 Plasma arc vitrification of incinerator ash *Proc. Int. Conf. Incineration and Thermal Treatment Technologies (Salt Lake City, UT, USA)*

[25] G.Z. Ye, E. Burstrom, M. Kuhn, J. Piret, Reduction of steel-making slags for recovery of valuable metals and oxide materials, *Scand. J. Metall.* 32 (2003) 7–14.

[26] R. Burkhard,W. Hoffelner, R.C. Eschenbach, Recycling of metals from waste with thermal plasma, Resour. Conserv. Recycl. 10 (1994) 11–16.

[27] Watanabe T and Shimbara S 2003 Halogenated hydrocarbon decomposition by steam thermal plasmas *High Temp. Mater. Process.* 7 455–74

[28] Z.G. Kostic, P.L. Stefanovic, P.B. Pavlovic, Comparative analysis of polychlorinated biphenyl decomposition processes in air or argon (plus oxygen) thermal plasma, J. Hazard. Mater. 75 (2000) 75–88.

[29] V. Ramasamy, T.X. Li, K. Saito, C.J. Cremers, V. Majidi, Combustion byproducts destruction by arc plasma, Hazard. *Waste Hazard. Mater.* 13 (1996) 131–141.

[30] H. Sekiguchi, T. Honda, A. Kanzawa, Thermal plasma decomposition of chlorofluorocarbons, *Plasma Chem. Plasma Process.* 13 (1993) 463–478.

[31] Wolf C. J., Heanley C. P., and Cashell P. V., (1998) Plasma arc system for vitrification of LLRW*Proc. 19th US Department of Energy Low-Level Radioactive Waste Management Conf.* (Salt Lake City, UT, USA).

[32] Carabin P., and Holcroft G., (2005) Plasma resource recovery technology—converting waste to energy and valuable products *Proc. 13th North American Waste to Energy Conf. (Orlando, FL, USA)* pp 71–9.

[33] Jouan A., Cartier R., Lannaud J., Guiberteau P., Durec J.P., Flament T., (1995), "Alpha waste management at the VALDUC research center", *International Conference on Evaluation of Emerging Nuclear Fuel Cycle Systems,* Versailles, France.

[34] Dmitriev S.A., Lifanov F.A., Stefanovsky S.V., Savkin A.E., Knyazev I.A., Polkanov M.A., Lashtchenova T.N., Golobokov S.M., (1997), *"High temperature treatment of solid low-level waste using shaft furnace-based plant with plasmafuel burners",* Waste Management 97, Tucson, USA.
[35] Kim T.W., Park J.K., Park S.C., Maeng S.J., SHIN S.W., (2001), "Non – combustible radioactive waste vitrification with plasma torch melter in Korea", *International Workshop on Plasma Processing for Nuclear Applications, Hanyang University,* Seoul, Korea.
[36] Barthelemy B., (2003), Thèse de doctorat, Combustion – Vitrification de déchets radioactifs par plasma d'arc : Modélisation de la thermique et de la dynamique Université de Limoges, France.
[37] Tzeng C.C., Hung T.M., (2001), "The status and prospects of development of thermal plasma processes for the treatment of hazardous wastes in INER", *International Workshop on Plasma Processing for Nuclear Application, Hanyang University,* Séoul, Corée, pp. 44-57.
[38] Hirabayashi T., Kanazawa K., Fujiki K., Yamate K., Ikenoya H., (1998), "Application of Plasma – Induction – Hybrid melter to the research on volume reduction and stabilization of low level radioactive solid waste", *International Conference on Incineration and Thermal Treatment Technologies,* Salt Lake City, USA, pp. 261-264.
[39] Amakawa T., Adachi K., Yasui S., Furukawa S., (2001), "Research and development of radioactive solid waste treatment technology by thermal plasma heating in Central Research Institute of Electric Power Industry (CRIEPI)", *International Workshop on Plasma Processing for Nuclear Application,* Hanyang University, Seoul, Korea.
[40] Adachi K., Furukawa S., Amakawa T., (2001), "Melting and solidification of surrogate TRU wastes by arc plasma", *15th International Symposium on Plasma Chemistry,* Orléans, France, pp. 2021-2027.
[41] Chateauvieux H., Guiberteau P., Longuet T., Lannaud J., Lorich M., (1998), "The VALDUC waste incineration facility starts active operations (IRIS process)", *International Conference on Incineration and Thermal Treatment technologies*, Salt Lake City, USA.
[42] Heberlein J., (1992) Generation of thermal and pseudothermal plasmas *Pure Appl. Chem.* 64 p 629.
[43] Westinghouse Plasma Systems (2006) http://www.westinghouse-plasma.com.
[44] Dighe S., (2007) *Westinghouse Plasma Systems, personal communication.*
[45] Hacala A., (2007) High power non transferred arc plasma torch applications *Proc. Thermal Plasma Fundamentals and Applications* (Sharm el Sheikh, Egypt).
[46] Hacala A., (2007) *Europlasma,* personal communication.
[47] Blary F., and Rollin M., (1995) Vitrification of asbestos wastes *Proc. Int. Symp. on Environmental Technologies: Plasma Systems and Applications (Georgia Tech Research Corp. Atlanta, GA)* pp 141–8.
[48] Chronopoulos C., Chevalier P., Picard J., Kaldas A., Carabin P., and Holcroft G., (2005) The plasma arc waste destruction system—one year of maritime experience *Proc. 24th Int.Conf. Incineration and Thermal Treatment Technologies (Galveston, TX, USA).*

[49] Carabin P., and Gagnon J-R., (2006) Thermal destruction of waste using plasma *Proc. Biomass and Waste to Energy Symp.(Venice, Italy)* (Italy: CISA Environmental Sanitary Engineering Centre).

[50] Holcroft G., (2005) A novel approach to waste treatment aboard ships and on land *Proc. 17th Int. Symp. Plasma Chemistry—Industrial Workshop (Toronto, Canada)* (University of Toronto).

[51] Kaldas A., Alexakis T., and Tsantrizos P., (2003) Optimization of the marine plasma waste destruction technology *Proc. IT3 Conf.* (Orlando, FL, USA).

[52] Ponelis A. A., and Walt I., (2003) Plasma conversion of fluorocarbon waste materials to useful products Proc. *16th Int. Symp. Plasma Chemistry* (Taormina, Italy) ed R d'Agostino et al (Bari, Italy: Universita di Bari).

[53] Bendix D., and Hebecker D., (2003) Energy recovery from waste and plasma conversion *High Temp. Mater. Process.* 7 435–54.

[54] Hanus G. J., and Springer M. D., (2007) Plasma thermal destruction and recovery treatment of PCB-contaminated liquids and solids *Proc. Int. Conf. Incineration and Thermal Treatment Technologies (Phoenix, AZ, USA).*

[55] Rutberg P. G., (2002) Some plasma environmental technologies developed in Russia *Plasma Sources Sci. Technol.* 11 A159–65.

[56] Rutberg P., Bratsev A. N., and Ufimtsev A. A., (2004) Plasmochemical technologies for processing of hydrocarbonic raw material with syngas production High Temp. *Mater. Process.* 8 433–45.

[57] Bratsev A. N., Popov S. V., Shtengel S. V., and Rutberg A. P., (2006) Some aspects of development and creation of plasma technology for solid waste gasification High Temp. *Mater. Process.* 10 549–56.

[58] Bacon M., (2005) Plasma assisted sludge oxidation Proc. 17th Int. S*ymp. Plasma Chemistry—Industrial Workshop* (Toronto, Canada) (University of Toronto).

[59] Chapman C. D., Deegan D. E., and Ly H., (2007) Integrated process solution for advanced thermal treatment of hazardous waste materials using plasma arc technology Proc. I*nt. Round Table on Thermal Plasma Fundamentals and Applications* (Sharm el Sheikh Egypt).

[60] Deegan D., Chapman C., and Bowen C., (2002) The production of shaped glass-ceramic materials from inorganic waste precursors using controlled atmospheric DC plasma vitrification and crystallization *Proc. 2002 EMRS Conf. (Strasbourg, France).*

[61] Chapman C. D., (2007) *Tetronics Ltd, personal communication.*

[62] Eschenbach R. C., (1989). Use of plasma torches for melting special metals and for destroying and stabilizing hazardous wastes. *Proc. Workshop Industrial. Plasma Applications* (Pugnochiuso, Italy) ed M I Boulos (Bari, Italy: Universita di Bari) p 127.

[63] Eschenbach R. C., Schlienger M. P., and Haun R. E., (1995) Swirl flow transferred plasma arc for vitrification of waste, metal recovery and special metal refining Proc. *Int. Symp. On Environmental Technologies: Plasma Systems and Applications* (Georgia Tech Research Corp., Atlanta, GA) pp 251–60.

[64] Eschenbach R C (1999) *RETECH personal communication.*

[65] F¨unfschilling M. R., Bernhard W., and Eschenbach R. C., (1992) Test results with the plasma centrifugal furnace Proc. *Plasma Technologies for Hazardous Waste Destruction* (Como, Italy) ed G Bonizzoni et al (Bologna, Italy: Editrice Compositori).

[66] Shuey M. W., and Ottmer P. P., (2006) LLW processing and operational experience using a plasma arc centrifugal treatemnt (PACT™) system *Proc. 2006 Waste Management Conf.* (Tucson, AZ, USA).

[67] Environmental Technology Evaluation Center (2002) *Environemntal Technology Verification Report for the Plasma Enhanced Melter* CERF/IIEC Report No 40633.

[68] Batdorf J., (2007) *Integrated Environmental Technologies, personal communication.*

[69] Carabin P., Palumbo E., and Alexakis T., (2004) Two state plasma gasification of waste Proc. *23rd Intern Conf. on Incineration and Thermal Treatment Technologies.*

[70] Psomopoulos C. S., Bourka A., Themelis N. J., (2009) Waste-to-energy: a review of the status and benefits in USA, *Waste Manage.* 29 1718–1724.

[71] Buragohain B., Mahanta P., Moholkar V.S., (2010) Biomass gasification for decentralized power generation: the Indian perspective, *Renew. Sust. Energy Rev.* 14 73–92.

[72] Minchener A. J., (2005) Coal gasification for advanced power generation, *Fuel* 84 2222–2235.

[73] Gomez E., Rani D.A., Cheeseman C.R., Deegan D., Wise M., Boccaccini A.R., (2009)Thermal plasma technology for the treatment of waste: a critical review, *J. Hazard.* Mater. 161 614–626.

[74] Byun Y., Namkung W., Cho M., Chung J.W., Kim Y.S., Lee J.H., Lee C.R., Hwang S.M., (2010) Demonstration of thermal plasma gasification/vitrification for municipal solid waste treatment, *Environ. Sci. Technol.* 44 6680–6684.

[75] Lee J.H., Kim Y.S., Do C.J., Hwang S.M., Jeong S.J., (2007) Hydrogen production by plasma gasification of waste, *J. Korean Soc. New Renew.* Energy 3 77–89.

[76] Sakai S., Hiraoka M., (2000) Municipal solid waste incinerator residue recycling by thermal processes, *Waste Manage.* 20 249–258.

[77] Moustakas K., Fatta D., Malamis S., Haralambous K., Loizidou M., (2005) Demonstration plasma gasification/vitrification system for effective hazardous waste treatment, *J. Hazard. Mater.* B123 120–126.

[78] Rani D.A., Gomez E., Boccaccini A.R., Hao L., Deegan D., Cheeseman C.R., (2008) Plasma treatment of air pollution control residues, *Waste Manage.* 28 1254–1262.

[79] Tzeng C.C., Kuo Y.Y., Huang T.F., Lin D.L., Yu Y.J., (1998) Treatment of radioactive wastes by plasma incineration and vitrification for final disposal, *J. Hazard. Mater.* 58 207– 220.

[80] Ghiloufi I., (2009) Simulation of radioelement volatility during the vitrification of radioactive wastes by arc plasma, *J. Hazard. Mater.* 163 136–142.

[81] Moustakas K., Xydis G., Malamis S., Haralambous K.-J., Loizidou M., (2009) Analysis of results from the operation of a pilot plasma gasification/vitrification unit for optimizing its performance, *J. Hazard. Mater.* 151 473–480.

[82] Jimbo H., (1996) Plasma melting and useful application of molten slag, *Waste Manage.* 16 417–422.

[83] Katou K., Asou T., Kurauchi Y., Sameshima R., (2001) Melting municipal solid waste incineration residue by plasmamelting furnace with a graphite electrode, *Thin Solid Films* 386 183–188.

[84] Nema S.K., Ganeshprasad K.S., (2002) Plasma pyrolysis of medical waste, *Curr. Sci. India* 83 271–278.

[85] Stiegel G., Ramezan M., (2006) Hydrogen from coal gasification: an economical pathway to a sustainable energy future, *Int. J. Coal Geol.* 65 173–190.

[86] Kim S.C., Chun Y.N., (2008) Production of hydrogen by partial oxidation with thermal plasma, *Renew. Energy* 33 1564–1569.

[87] Galvita V., Messerle V.E., Ustimenko A.B., (2007) Hydrogen production by coal plasma gasification for fuel cell technology, *Int. J. Hydrogen Energy* 32 3899–3906.

[88] Byun Y., et al. (2011) *Journal of Hazardous Materials* 190 317–323

[89] Ladebeck J.R., Wagner J.P., (2003) Catalyst development for water–gas shift, in:W. Vielstich, A. Lamm, H.A. Gasteiger (Eds.*), Handbook of Fuel Cells—Fundamentals, Technology and Applications,* John Wiley & Sons, Chichester,.

[90] Lv P., Yuan Z., Ma L., Wu C., Chen Y., Zhu J., (2007) Hydrogen-rich gas production from biomass air and oxygen/steam gasification in a downdraft gasifier, *Renew. Energy* 32 2173–2185.

[91] Cao Y., Wang Y., Riley J.T., Pan W.P., (2006) A novel biomass air gasification process for producing tar-free higher heating value fuel gas, *Fuel Process. Technol.* 87 343–353.

[92] Compoy M., Gómez-Barea A., Vidal F.B., Ollero P., (2009) Air-steam gasification of biomass in a fluidized bed: process optimization by enriched air, *Fuel Process. Technol.* 90 677–685.

[93] Li X.T., Grace J.R., Lim C.J., Watkinson A.P., Chen H.P., Kim J.R., (2004) Biomass gasification in a circulating fluidized bed, *Biomass Bioenergy* 26 171–193.

[94] Colombel P., (1996) Étude du comportement à long terme de vitrifiats de REFIOM. *PhD thesis,* France: Université de Poitiers.

[95] Vernaz E., (1993) Étude du comportement à long terme des matériaux vitrifiés, *Lecture presented at POLLUTEC industrial trade meeting.*

[96] Colombel P., Godon N., Vernaz E., Thomassin JH., (1995) Mécanismes d'altération des vitrifiats de REFIOM et analogie avec d'autres verres industriels ou naturel. In: *Proceedings of the International Conference on Waste Solidification-Stabilization Processes,* Nancy, France.

[97] Frugier P., Godona N., Vernaza E., Larchéb F., (2002) *Influence of composition variations on the initial alteration rate of vitrified domestic waste incineration fly-ash Waste Management* Volume 22, Issue 2, p 137-142.

[98] Le Forestier L., (1996) Résidus ultimes de l'incinération de déchets ménagers. *PhD thesis,* France: Institut National Polytechnique de Lorraine.

[99] Frugier P., (1999) Altération des vitrifiats de REFIOM: Influence des variations de composition et étude du terme source. *PhD thesis,* Université de Montpellier.

[100] Fautré R. (1998) La vitrification par aré álectrique d'un doéchet industriel: le REFIOM, Société Française de Minéralogie et de Cristallographie, *Journées d'étude sur "Les verres: des matériaux pour le troisième millénaire".*

[101] Tippayawong N., Khongkrapan P., (2009) Development of a laboratory scale air plasma torch and its application to electronic waste treatment *Int. J. Environ. Sci. Tech.,* 6 (3), 407-414.

In: E-Waste: Management, Types and Challenges
Editors: Yuan Chun Li and Banci Lian Wang
ISBN 978-1-61942-217-9

Chapter 9

AN OVERVIEW OF TOXIC ENVIRONMENTAL RELEASES FROM E-WASTE

M. Mosharraf Hossain,[1] Ismail M. M. Rahman,[2,3,*] Zinnat A. Begum[3] and Hiroshi Hasegawa[3]

[1]Institute of Forestry and Environmental Sciences, University of Chittagong, Chittagong, Bangladesh
[2]Department of Chemistry, Faculty of Science, University of Chittagong, Chittagong, Bangladesh
[3]Graduate School of Natural Science and Technology, Kanazawa University, Kakuma, Kanazawa, Japan

ABSTRACT

Electronic wastes have become the foremost stream of solid waste in almost all the countries of the world over a short span of last few years. The digital information and communication technology (ICT) revolution that the mankind is undergoing has substantially contributed in creating this scenario. Popularly abbreviated as e-wastes, electronic waste comprises of all out-of-date discarded or otherwise unused electronics and peripherals thereof – the bulk being computers of many descriptions, consumer electronics and mobile phones. Necessarily, a huge amount of electronics is replaced by individuals and organizations each year but people are increasingly purchasing more electronics than they really need with the rapid introduction of cheaper, sleeker and alluring gadgets. The technology and trends are changing quite rapidly rendering usable electronics obsolete and make us to go for newer versions of the same thing thereby adding up to the volume of outmode electronics. Bulk of these ends up into landfills, for USA 80-85% of e-wastes goes to landfills as estimated by US-EPA. Volume of e-wastes is a big concern but the high content of heavy metals, *e.g.* lead, chromium, cadmium and mercury, etc., along with harmful organic compounds in the electronic products is a bigger concern for mankind. To some, the name 'e-waste' is ironic as it contains over 60% recyclables including iron, copper, aluminum, gold and other metals let alone the polymers from casings of disposed electronic products. The increasing flow of e-wastes naturally stimulated various solution responses, one of which is the boom in e-waste

* Author for correspondence. *E-mail:* I.M.M.Rahman@gmail.com.

recycling business, in the form of formal industry in developed part of the world while manual and archaic disassembly of wastes in developing countries most of which are imported from developed countries for recycling. The non-recyclable parts of these wastes, including the toxicants, are finding their way into the environment and it has led to the formulation of Basel Ban, an amendment to the Basel Convention to restrict trans-boundary movement of e-waste. However, the e-waste flux from developed to developing countries is on the rise due to win-win economics of informal recycling. Our knowledge about the chemistry of distribution, speciation, transport and environmental fate of these hazardous elements in and outside landfills are still sketchy though some researchers have initiated in-depth studies to assess and predict the eco-environmental consequences of the toxicants from disposed e-wastes. The incineration at the informal reprocessing sites contaminates the air and soils with polychlorinated and polybrominated dibenzo-*p*-dioxins and dibenzofurans, polycyclic aromatic hydrocarbons, etc. On the other hand, acid leaching processes contaminate the freshwater sources and sediments with heavy metal. The effects of these releases are exacerbated through the entry of pollutants into the food cycle. Besides the contamination of the surroundings, the recycling process itself poses health risks to the workers out of exposure to the chemicals used in recycling and hazardous byproducts from the recycled wastes. Within the scope of this chapter, we will provide a brief overview, in light of the recent findings, of the toxic environmental releases from the e-waste disposal and recycling processes and their effects.

INTRODUCTION

E-waste or electronic waste, which includes waste electrical and electronic equipment (WEEE), is a growing concern all over the world [1], primarily because of the exponential increase in their types, volumes and sources [2]. Adding to it is the decreasing life-span of electronic products due to fast-paced innovations [3] and specially, due to the environmental [4] and health hazards [5, 6] these wastes may pose. According to a 2005 estimate [7], every year 20–50 million tons of WEEE is generated worldwide with an annual growth rate of 3–5%, three times faster than the growth of the solid waste stream. On the other hand, according to the 2010 estimate [8], the total volume of global electrical and electronic products' input to the market was 57 million tons with the projection for 2015 at 76 million tons. Alongside the dwindling resource base, upon which the future growth of the electronic industry depends, the industry requires to reuse and recycle the materials from end-of-life (EoL) electronics. It has created a positive impetus towards sustainable e-waste management [9]. However, the management capability as well as the awareness towards the e-waste issue differs among different nooks of the world [10] - the developed nations can bear the e-waste management cost due to better awareness among the people and are capable of coming up with the technological intervention needed for e-waste management unlike the developing and underdeveloped portions of the world [11]. This has created a disparity between these two segments of world's countries in terms of hazards they are getting exposed to. This is critical since the countries belong to the latter region are not the producers of electronics – they are the consumers of products from developed world and are the worst sufferers of problems put forth by the e-waste issue. There is an increasing global effort to fight the issue of e-wastes. Organizations like the United Nations Environment Programme (UNEP), the Secretariat of the Basel Convention and StEP (Solving the E-waste Problem) - an initiative developed by a number of United Nations organizations, have taken the lead in providing guidelines and assistances in the form of regional and international conventions or directives. As the stream of e-waste is increasing and there are multitudes of approaches in fighting this concern, a particular worry rose by different practical examples and studies from many parts of the world involves the toxicity pertaining to e-waste and their recycling.

In this review, we have evaluated the issues related to the release of hazardous substances from e-waste under the recycling scenario as well as in the non-management scenario where e-wastes are dumped into the landfill sites as a part of the solid waste stream. In the *first section*, we have tried to figure out the categories of e-waste the world is producing now and the major contributors at present with a future projection of possible big contributors. This section will also focus on the content of e-wastes in order to identify the potential harmful substances in them which get released upon their disposal or during recycling processes. The *second section* has given a detail picture of toxic environmental releases from the e-wastes as recorded in available literatures from different parts of the world along with the associated environmental and health issues. The *third section* has focused on the geographical movement of e-waste, and the factors related to that in order to see the play of economics and politics in the issue of toxic discharges during the processing of e-wastes. The *fourth section* has addressed diverse aspects of e-waste management pertaining to the control of hazardous releases during recycling process, or once it gets into the environment. We shall finish this review by drawing a comparative discussion on e-waste management experiences in different parts of the world with particular emphasis on the toxic releases in the *fifth and final section*. In the conclusion of the review, we have included the factors needed to be focused in e-waste management to check or address the environmental toxicity issues.

E-WASTES: CATEGORIES AND RELEASABLE ENVIRONMENTAL TOXICANTS

Electrical and electronic devices are manufactured using a complex blend of components, which often contain several hundred different substances - both toxic and non-toxic in nature. E-wastes includes large household appliances (*e.g.* TV, refrigerators, air-conditioning systems, washing machines, etc.), small household appliances, IT and telecommunication devices, lighting equipment, leisure and sports equipment, medical devices, monitor and control instruments, automatic dispensers, hand-held digital apparatuses and so forth. Some of these have a life-cycle of only a few years or even less. E-waste, in most of the cases, contain many hazardous substances - extremely dangerous to human health and the environment, and therefore, their disposal requires special treatment to prevent leakage and dissipation of toxics into the environment [12]. More *al*'s [13] overview on the chemical composition of different electronic parts from WEEE generated in Switzerland gives a fair idea about the content in e-waste in other parts of the world (Table 1).

In general, electronic equipment are made of a considerable quantity of plastic components containing toxic trace metals which are added as catalysts, stabilizers or colorants [14]. Integrated circuits (ICs) in most of the scrap electronic products contain various metals such as iron (Fe), aluminium (Al), zinc (Zn), gold (Au), silver (Ag), copper (Cu), nickel (Ni), lead (Pb), mercury (Hg), cadmium (Cd) and chromium (Cr). On the other hand, each computer screen contains about 20% lead by weight. Traditional cathode ray tubes

(CRT) are waning due to the environmental concerns, yet the newer generations of displays like plasma and liquid crystal display (LCD) creates human health toxicity as they contain antimony (Sb), arsenic (As), barium (Ba), beryllium (Be), cadmium (Cd), chromium (Cr), cobalt (Co), copper (Cu), lead (Pb), mercury (Hg), molybdenum (Mo), nickel (Ni), selenium (Se), silver (Ag), vanadium (V), and zinc (Zn). However, the new devices are more threatening in terms of ecological toxicity caused from Hg used in LCD displays and Cu used in plasma displays [15]. Pb, Hg, Cd and Cr are known environmental pollutants, and they are included in the list of harmful substances of RoHS directive which requires the manufacturers of electronic products to exclude these metals in future electrical and electronic products in Europe. According to RoHS diredctive, the maximum tolerable mass fractions are 0.1% of Pb, Hg, Cr^{6+}, PBB and PBDE, and 0.01% of Cd in homogeneous materials [16]. Many countries are now trying to conform to this directive. Besides these heavy metals, other hazardous substances included in e-waste are brominated flame retardants (BFR) such as polybrominated biphenyls (PBB), polybrominateddiphenylethers (PBDEs) [7], hexabromocyclododecane (HBCD), ethylene(bis-tetrabromophthal)imide (EBTBPI) and bis (pentabromophenyl)ethane (BPBPE) etc.

Table 1. Contents of selected metals and non-metals in Swiss-WEEE

Element/substance	*Average value±Range of uncertainty (mg kg^{-1})*
Al	49000±4000
Sb	1700± 200
Pb	2900±500
Cd	180±50
Cr	9900±1700
Fe	360000±10,000
Cu	41000±3000
Ni	10300±1600
Hg	0.68±0.22
Zn	5100±800
Sn	2400±100
Cl	9600±1400
P	360±220
PCB Sum	13±4

The waste plastic casing of mobile phones reportedly [17] contains 5.0–340 mg kg^{-1} of Pb, 4.6–1005 mg kg^{-1} of Cd, 5.0–11000 mg kg^{-1} of Ni and 5.0–12500 mg kg^{-1} of Ag depending on makers and models. It clearly represent the potential threat of toxic environmental releases from this category of waste given the short life span of mobile devices and the increasing tele-densities all over the world. In another study on the metallic content of EoL mobile devices [18], more than twenty metal elements have been reported out of which twelve were highly hazardous, and they constituted 35–40% of the total weight of devices.

The toxic release of persistent organic pollutants (POPs) to different media of environment has been reviewed [19] at Guiyu, China–an e-waste processing area, and a higher concentration of POPs were found in air, soils and water compared to the control sites. The blame for toxic environmental releases goes not only to the e-wastes but also to the chemicals, for example, strong acids and organic solvents, which are used in extracting metals and polymers from waste materials.

ENVIRONMENTAL AND HEALTH HAZARDS OF TOXIC RELEASES FROM E-WASTE

Part of e-wastes ends up, in most of the countries, in landfills from where the leakage of chemical and metal alongside the vaporization of metallic mercury and dimethylene mercury creates health hazards [7]. Leaching of other heavy metals, through natural and biological processes, is also becoming a worrying concern with the increasing percentage of e-wastes in the municipal solid waste stream. Furthermore, the polymers of diverse kind and character, as mentioned earlier, can't be ignored in terms of the potential leaching of these through the microbial breakdown, into the water and soil of landfill sites, which may pose a very long-term health risk.

An excellent review [20] on the biological effects of polybrominateddiphenyl ethers (PBDEs), tetrabromobisphenol A (TBBPA) and derivates, hexabromocyclododecane (HBCD) and polybrominated biphenyls (PBBs) give a scary picture of diverse health issues related to e-waste mismanagement on both human and other animals. PentaBDEs are toxic at comparably lower doses than decaBDEs. PentaBDEs affect neurobehavioural development and thyroid hormone levels in rats and mice, octaBDEs cause teratogenicity in rats and rabbits while decaBDEs affect thyroid, liver and kidney morphology in adult animals. HBCD has an effect on the liver and on thyroid hormone; PBBs are reported to have effect on reproduction and carcinogenicity alongside its role in neurological and musculoskeletal problems. Another review [21] has specifically highlighted the harmful role of BFRs in disrupting endocrine pathways of thyroid hormones, endrogen and estrogen along with neural development.

Level of Pb in blood and the frequencies of lymphocytic micronucleated binucleated cells (MNBNCs) were higher in workers exposed to higher level of atmospheric Pb in e-waste processing facility in China compared to the control group [22]. This observation indicated the health risk associated with occupational environments at e-waste processing sites and shows the inevitability of containment of hazardous materials to ensure risk-free management of e-waste. A study from China [23] reported significantly higher ($p<0.01$) blood-lead levels (>10 μg dL^{-1}) in 81.8% of the Children living in Guiyu – the e-waste recycling area, compared to 37.7% in those living in the control area, Chendian. The bioaccumulation of PBDEs in the form of BDE47 was observed in the fish - Chinese loach (*Misgurnusanguillicaudatus*) – cultivated in a net-cage at an e-wastes recycling site [24], which indicates a bigger concern of assimilation of toxic environmental releases into the food chain. Report from Vietnam [25] gives a graver indication of food-cycle accumulation of toxic hazards from e-waste through human breast milk; recyclers from Bui Dau area showed the significantly higher levels of PBDEs (20–250 ng g^{-1} lipid wt.) and HBCDs (1.4–7.6 ng g^{-1} lipid wt.) in breast milk compared to control group. Similar results have been reported from China [26] and Australia [27] as well. A specific accumulation, unrelated to diet, of low-chlorinated PCBs and high-brominated PBDEs was observed [25] in e-waste recyclers, suggesting extensive exposure to these compounds during e-waste recycling activities, possibly through inhalation and ingestion of dust. Investigations into the health cost of e-waste are also underway in India [28].

The soils of e-waste processing areas are prone to the heavy-metal contamination which affects the soil productivity such as supporting paddy growth since the germination rate, and the root lengths were negatively affected by the metal contamination [29].

The Geographical Perspective of E-Waste Economy

Global annual production of e-waste, in 2009, was estimated at 20–25 million tones [30]. At present, Europe, the United States and Australasia are generating the bulk of this quantity. However, China, India, Eastern Europe and Latin America will become the major player in the years to come. On the contrary, the trans-boundary movement of e-wastes, often violating the Basel ban, has so far created informal low-tech e-waste processing facilities in developing and underdeveloped parts of the world. Among which biomedical concerns are documented in countries, including China, India, Vietnam, Nigeria, etc. [1, 26]. Even under such a scenario only 10% of the world's e-waste can be recycled [31].

Economic Consequences of E-Waste Management

From a theoretical perspective, in case of e-waste management, a production theory based material flow analysis scheme that includes the accounting for material back flows from recycling of e-wastes has been put forward by Schmidt [3] by using Hewlett-Packard Europe as a case. Theory of business and management economics and the concept of material flow nets from environmental informatics have been used in the study. Classical production system does not care for waste, as it had not been a big issue since recent past. However, in traditional systems, inclusion of e-waste recycling component only focuses on the impacts on the environment whereas the ideal e-waste management should include the both. Some companies from industrialized countries opt for a harmful but cost-savvy and sometimes illegal waste management option of exporting the waste to developing countries as second hand or refurbished products. They do so to take the advantage of lower purchasing power, unawareness, poor policy and legislation framework and cheaper labor cost but in these ways they are exposing the population of underdeveloped countries to unprecedented environmental and health hazards.

Political Economy of E-Waste

E-wastes have long been exported to the developing and underdeveloped countries of the world. The issue of movement of these hazardous materials capable of toxic environmental releases was promoted due to the lack of awareness in the developing part on one hand and the economic and political pressures, on the other hand. Though Basel Ban aims at inhibiting such movements, considerable quantities of e-waste are exported as used and refurbished electronics as well as raw material for formal and informal recycling businesses in developing and developed countries of the world. The corrupt political system in the developing and developed countries is exploiting the lack in environmental awareness among the general mass to sustain or to promote this business. The issues related to such movement of e-wastes have been dealt in detail in the reviews by Widmer *et al.* [4], Ongondo *et al.* [10], Wong *et al.* [19], LaDou & Lovegrove [31], Nnorom & Osibanjo [32], Rosenfeld & Feng [33], etc. The traffic of e-waste in the Asian region is given in Figure 1 as adapted from Widmer *et al.* [4].

MANAGEMENT OF E-WASTE

The e-waste recycling chain comprises of three main steps *viz.* collection, preprocessing (sorting, dismantling and mechanical treatment) and end-processing–the holistic performance of these three ensures the success of the overall recycling programs. According to UNEP's Solving E-waste Problem (StEP) initiative [34] - the sustainable management of e-waste aims to achieve the following objectives –environment-friendly treatment of the hazardous part of e-waste, maximizing the recovery of valuable materials from e-waste for promoting eco-friendly business along with due consideration for local and social contexts. E-waste is a special class of waste, and it leads to the development of new principles and policies for its management while it gives rise to management challenges in terms of recycling and reuse policies, recycling technology, economic burden and environmental consequences. For example, PCB and Hg, both are found in condensers, accumulators, batteries and mercury switches, etc.; their separate collection during the e-waste treatment process was proven very effective in avoiding contamination of output fractions with these hazardous substances while hand-sorting or mechanical separation was proven ineffective in collecting Cd, Pb, Sb and these may become a hazard to humans and the environment [13].

Principles and Policies

In response to the issue of e-waste management challenges, some policies and programs are becoming widely adopted among many countries which aim mainly at pollution prevention and waste minimization. Some of those approaches are discussed below:

Extended Producer Responsibility (EPR)

The EPR recognizes that the manufactures are financially responsible for the entire life-cycle of their products even after their products reach EoL. The approach deals with one basic assumption that the manufacturers know the best mean of dismantling the products that they have designed. Furthermore, as the designer and manufacturer of the product they are capable of efficient reuse and recycling of the recovered materials from e-wastes. Such approaches are in place in Europe, Japan, Taiwan, etc. under some legal framework and China, India and other developing countries are also gradually moving towards that direction.

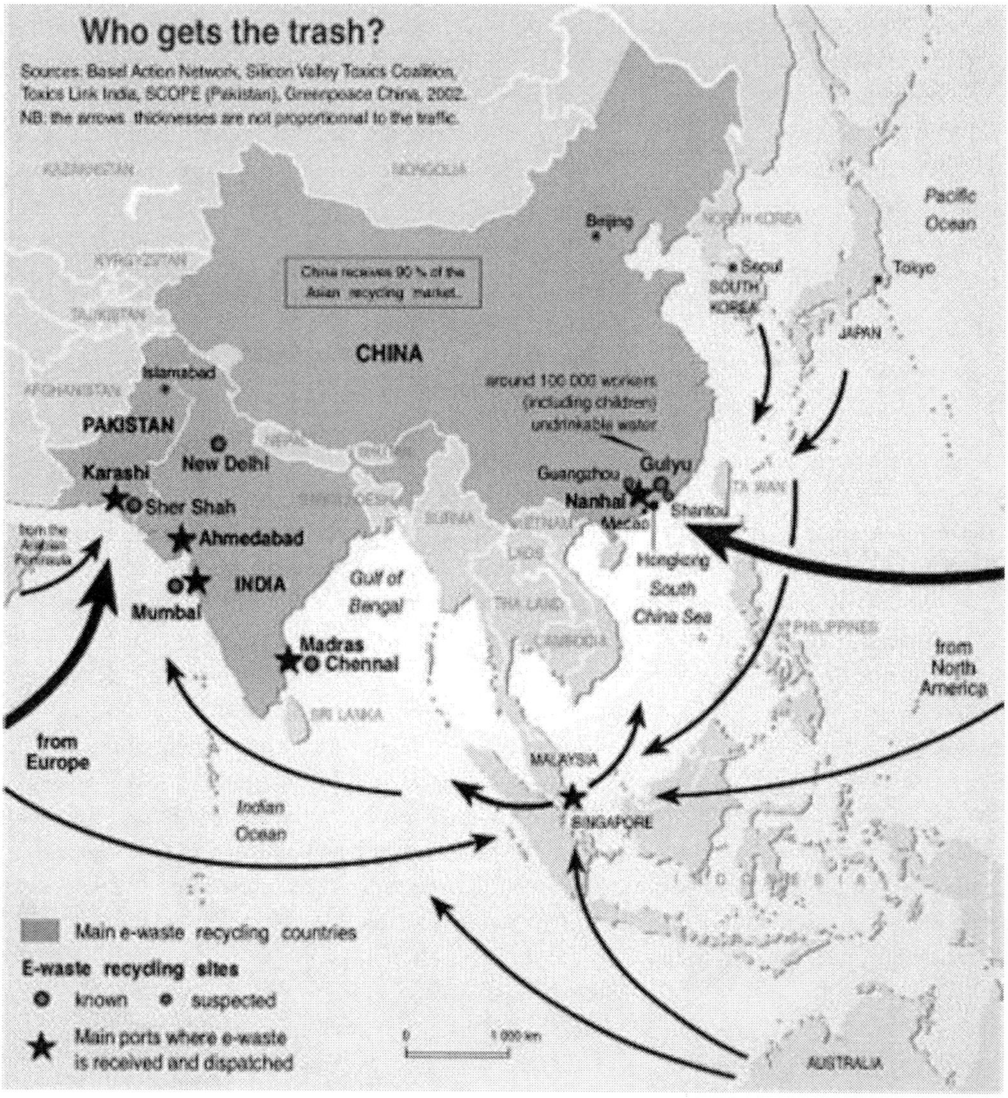

Figure 1. Asian e-waste traffic.

"Reprinted from *Environmental Impact Assessment Review*, 25/5, R. Widmer, H. Oswald-Krapf, D. Sinha-Khetriwal, M. Schnellmann and H. Böni, Global perspectives on e-waste, p. 436–458, ©2005, with permission from Elsevier."

Redesign of Electrical and Electronic Products to Make them Recycle Friendly

In this approach, the manufacturers study the e-waste problems created by their products and the hurdles that the current design of the products pose on the management of these products at EoL. The results are then used to redesign their products to avoid potentially hazardous materials as much as possible while designing the products for easier recycling. Legislative measures, such as the Restriction of the Use of Certain Hazardous Substances (RoHS), is an example of policy push towards recycle friendly less-hazardous e-waste. The RoHS directive includes mainly six substances *viz.* lead (Pb), mercury (Hg), cadmium (Cd), hexavalent chromium (Cr^{6+}), polybrominated biphenyls (PBB), polybrominateddiphenyl ether

(PBDE). Due to the implementation of RoHS directives, the printed circuit board industry replaced lead containing solders and surface finishes with alternative materials like organic solderability preservatives, electroless nickel/immersion gold, immersion silver and immersion tin, which resulted in the decrease of global market share of lead containing PCBs from 55% in 2004 to a staggering 37% in 2004 [8]. In this context, we need to adopt newer approaches like life cycle analysis (LCA) [35] for being able to come up with better material flow and superior process design in managing the toxic environmental and health hazards from e-waste.

Technological Advancements and Bottlenecks in E-Waste Management

There are many options so far developed and new approaches are in the pipeline for addressing the recycling issue of metal and non-metal [36] from e-wastes. The intensive research is underway since the industries are now held responsible for wastes that their products may release to the environment alongside the necessity of material recovery for the sustainable production.

Detection

Detection of the presence of releasable but trace amount of toxic substances in e-waste is quite important. It is very challenging and an active area of current research. There have been some positive developments coming in this area of e-waste management. Ion attachment mass spectrometry (IAMS) based rapid and sensitive method has been developed to test polymers from e-wastes for the presence of polybrominated-diphenyl ethers (PBDEs). It did not require any chemical pretreatment or separation process for the individual organic compounds before analysis, and the method successfully determined the presence of decaBDE - a BFR, in polymer at a concentration as low as 13.5 ppm per mg of solid sample. Even when brominated compounds, such as ethylene (bis-tetrabromophthal)-imide (EBTBPI) and bis-(pentabromophenyl)-ethane (BPBPE) were present in the sample together with decaBDE, IAMS successfully identified each [37]. On the other hand, recently Durate *et al.,* [14] has reported direct solid sample graphite furnace atomic absorption spectrometry without sample pretreatment to detect the presence of Cd and Pb in reference plastic materials from e-wastes with a detection limit of 0.1 mg kg^{-1} for Cd and 0.6 mg kg^{-1} for Pb, a satisfactory agreement with the reference values was also confirmed. Direct exposure probe/mass spectrometry (DEP/MS) was investigated in the detection of deca-BDE with a limit of detection (LOD) and limit of quantitation (LOQ) of 0.163 and 0.495 mg kg^{-1}, respectively [38]. In order to get a detail picture of the available techniques to determine BFRs especially PBDEs in environmental and human samples, the review by Covaci *et al.* [39] can be consulted.

Recycling and Recovery

Modern recycling plants can recover 80% of the material from the scrap and use another 15% for burning while only 5% finishes as waste. Lee *et al.,* [40] reported the removal of plastic resin sealing of scrap ICs by roasting the samples at 850°C, followed by screening and magnetic separation and then the recovery of Au, Ag and Cu by leaching with sulfuric acid, ammonia, aqua regia and thiourea. A 100% recovery could be attained by thiourea. The finding can lead to the automated industrial setup for recovery of these valuable metals for reuse. On the other hand, Sato *et al.* [41] reported 81.4% and 85.4% recovery of the decaBDE, a BFR contained by polymer materials from e-waste, respectively, for the samples containing 317-ppm and 886-ppm of decaBDE per mg of solid polymer sample. Under different conditions, the efficacy of pressurized liquid extraction (PLE), microwave-assisted extraction (MAE), and ultrasonic-assisted extraction (UAE) of BFR from the polystyrene matrix have been investigated [42]. It was seen that PLE could extract 95–100% of TBBPA and HBCD, and 50% of deca-BDE while MAE gave lower extraction yields for TBBPA and UAE gave poor (10–50%) recovery rates. Zhong and Peng [43] studied the effect of the presence of different metals on the organic solvent-based (*i.e.* acetone, methanol or toluene) removal or recovery of hexabromocyclododecane (HBCD), a widely used BFR and concluded that the presence of Ni enhances the transformation of HBCD (45–99%) significantly while other metals have negligible effect on the extraction process. They have suggested the exclusion of acetone in the recycling or extraction of HBCD from Ni-rich e-wastes owing to the elevated risk of debromination of HBCD even at mild extraction temperatures.

Incineration

Incineration is one of the options of e-waste management, which increases the risk of toxic environmental release. There is a risk of dioxins, furans, polycyclic aromatic hydrocarbons (PAHs), polyhalogenated aromatic hydrocarbons (PHAHs), and hydrogen chloride generation from incinerating e-waste [30]. Incineration of e-waste at informal recycling sites contaminates air and soils with polychlorinated and polybrominated dibenzo-*p*-dioxins and dibenzofurans, polycyclic aromatic hydrocarbons (PAHs) etc. Further details on the incineration issue can be seen in the comprehensive review by Tsydenova and Bengtsson [44].

Solvent Based Extraction

Acid leaching processes are widely used in recovery of materials from e-wastes. In this process, the environmental releases from e-wastes contaminate the freshwater sources and sediments with heavy metal, which is exacerbated through the entry of these pollutants into the food cycle. Some alternatives to acid leaching have already been proposed. Nnoroma and Osibanjo [45] reviewed environmentally sound management options for waste plastics from electronic wastes in Nigeria where they have suggested that the Creasolv® and Centrevap® - two solvent-based approaches - developed through extensive research on management of waste plastics from WEEE in Germany and United Kingdom. Creasolv® process could remove most of BFRs including dioxins and furans from styrenic polymers whereas Centrevap® was more successful in removing other submicron insoluble impurities from a wide range of polymers.

Bioleaching

We also need to look for microbes, which are capable of leaching metals and non-metals from waste electronics. For example, *Chromobacterium violaceum* was successfully tried [46] in leaching out 11.3% of Au and 24.6% of Cu from waste mobile phone PCBs in an environmental friendly manner. Brandl and Faramarzi [47] reported microbial mobilization of

Cd, Cu & Zn (>80%), Pb (2%), Cr (11%) and Ni (32%) by *Acidithiobacilli* from particulate fly ash – the remnant of incineration of e-waste containing solid waste. They also have effectively used *Chromobacterium violaceum* in mobilizing Ni and Au from electronic waste. Accordingly other metal leaching species of bacterium or fungi can be tried for better recovery of metals from e-wastes at low-cost and in an environment-friendly manner.

Comparative Scenario in E-Waste Management among the Countries

Switzerland was the first country to implement an industry-wide organized system for the collection and recycling of electronic waste [12]. A comparative review [48] on the e-waste recycling issue covered the systems for e-waste recycling in Europe, Japan, South Korea, Taiwan and USA. The EU's WEEE Directive requires the electronic and electrical products' manufacturers to take back EoL equipment, free of charge, for environmentally sound disposal. The directive also specifies recycling quotas in order to reuse and channel back the old appliances or their materials into the economic cycle to ensure a linear material flow management in order to convert e-waste management from a throw-away mentality type management into a closed-loop material flow management with a high recycling target [3]. WEEE directive initially set the target of e-waste collection per capita per annum at four kilograms, but later they have corrected it by giving the directive to collect 65% of the average weight of products placed on the market by the producer in the two preceding years, which will become a binding requirement in 2016 [49]. On the other hand, the RoHS directive enacted by Europe and adopted in some parts of USA (*e.g.* California) and in other countries (*e.g.* Canada, Japan, China, Korea and Turkey). However, USA has not yet adopted legally enforceable federal policies for regulating e-waste though examples of sporadic efforts are there [1]. China is becoming the world's largest manufacturer of electrical and electronic products but a recent study [50] presented a grim picture of their adoption and implementation of e-waste management initiatives, though they have already put legislative measures such as China WEEE and China RoHS. The Chinese versions of WEEE and RoHS directives in place since 2005 and 2007, respectively. As of July 1997, the environmental protection agency of Taiwan declared scrap personal computers is the producer's recycling responsibility.

REACH (Registration, Evaluation, Authorization and Restriction of Chemicals) is another European Union Regulation to force the use of potentially toxic substances (such as phthalates or BFRs) to be phased out from electronic and other products. It is the strictest law to date regulating chemical substances requiring all companies manufacturing or importing products containing chemical substances into the European Union in quantities of one ton or more per annum to register these substances with a new European Chemicals Agency (ECHA) based in Helsinki, Finland. The regulation entered into force on June 1, 2007 with a phased implementation over the next decade [51-53].

In China and India, a license scheme for e-waste recycling was started [54] when the simple prohibition on informal and improper recycling failed to control hazardous recycling but the success of such licensing efforts was seen to be dependent on the responsibility of e-waste disposer. In Germany, the recycling industry dismantles and recycles electrical equipment chiefly in workshops by handicapped or challenged persons [3], which is an example of the involvement of the social component in the e-waste recycling sector.

Issues Need to Be Focused in E-Waste Management

Alike other researchers [12], we agree that there is no singular panacea to address the e-waste issue, and we need to find optimal and contextual solutions for specific cases depending on the technological, economic and cultural realities of a particular nation or locality. However, there should be some general guiding principle and set targets to measure the efficiency of e-waste management programs – both in the public and private sectors. The universal directives and standards are on the process of formulation – the examples being Basal convention, StEP, WEEE, REACH and RoHS directives – but we still have a long way to go to ensure that the toxic release or the chance of such a release from e-waste is brought down to a minimum. We need to look at the emerging trends in order to do so – some of the issues are sketched below.

In the e-waste management and directives so far, the developed part of the world and their industrial sector has been included. However, in recent time, we can see that the developing countries have started their own electrical and electronic (EE) industries, which are expected to take off at a rapid pace due to the cost and labor advantages along with the poor legislative and monitoring framework for overseeing the hazards they put forward. With the advent of such low cost electronic manufacturers to remain at par with the buying capacity of the low-income consumers, there will be a tendency to bypass the international conventions and directives regarding the use of greener materials, better designs, and hence it will aggravate not only the waste generation by volume but also the hazardous content in the e-waste stream in these countries. In any future conventions, therefore, this issue is required to be considered for better legal and technological cooperation worldwide ensuring the due and active participation of these nations with potentially booming EE industries. Moreover, a clearer modality should be in place to objectively estimate the e-waste generations, both quantitatively and qualitatively, on a global scale. It can be ensured through the mandatory participation of the manufacturers in building a regularly updated database, managed, for example, by UNEP, where the outputs of EE products and the ingredients used therein should be updated by the manufacturers themselves. In this age of globalization, another issue needs attention is that, in the developing countries, where usually the EE products are mainly imported, proper implementation of the EPR principle is difficult and so does the recycling by the manufacturer of these products. We need to develop alternative-waste management principles in these countries, where an extended cooperation between the local manufacturers or recyclers and the original producers for combating e-waste will be needed. In many developing countries, the e-wastes are managed by informal sector, which maximizes the risk of human exposure to toxic releases. In such cases educating the unofficial sector of the risks, precautions to be taken against those risks and their management once exposed is another important issue that needs consideration. Under the current scenario of e-waste management, most of the countries are following a 'multiple enforcement body' approach, which is to be made centralized for the simpler control and quick response on the part of the authority and easier compliance assurance on the part of the industry.

CONCLUSION

We conclude from the facts and developments in the issue of toxic releases from e-waste that, the risk of such occurrences, even when many initiatives and directives are in place, are on the rise with the rapidly increasing flow of e-waste due to the advancement on technology and changes in our habits. There are tremendous environmental and health consequences of such releases as outlined in a vast number of works by researchers from different parts of the world. The initiatives are in place but there are many factors that hinder the proper translation of those initiatives into realities, particularly in the developing and underdeveloped part of the world. In order to make the electrical and electronic industries more sustainable, we have to pursue more research towards the environment friendly and recycle-prone products design. We have to change the attitude of producers and consumers towards producing or using long life-span items instead of short life span gadgets. Moreover, the waste-management options based on 3Rs are to be enhanced by developing better extraction and recycling options at lower costs. A complete ban over the use of hazardous and potentially toxic materials in manufacturing electronic products are to be ensured, which will minimize the burden on the management of such wastes. In order to ensure a safe use and benign disposal of e-wastes by avoiding the potent lethal environmental releases, global-scale collaboration is eminent instead of segmented approaches.

REFERENCES

[1] Ogunseitan, O. A.; Schoenung, J. M.; Saphores, J.-D. M.; Shapiro, A. A., The electronics revolution: From e-wonderland to e-wasteland. *Science.* 2009, *326*, 670–671.

[2] Babu, B. R.; Parande, A. K.; Basha, C. A., Electrical and electronic waste: A global environmental problem. *Waste Manag. Res.* 2007, *25*, 307–318.

[3] Schmidt, M., A production-theory-based framework for analysing recycling systems in the e-waste sector. *Environ. Impact Asses.* 2005, *25*, 505–524.

[4] Widmer, R.; Oswald-Krapf, H.; Sinha-Khetriwal, D.; Schnellmann, M.; Böni, H., Global perspectives on e-waste. *Environ. Impact Asses.* 2005, *25*, 436–458.

[5] Frazzoli, C.; Orisakwe, O. E.; Dragone, R.; Mantovani, A., Diagnostic health risk assessment of electronic waste on the general population in developing countries' scenarios. *Environ. Impact Asses.* 2010, *30*, 388–399.

[6] Chen, A.; Dietrich, K. N.; Huo, X.; Ho, S. M., Developmental neurotoxicants in e-waste: An emerging health concern. *Environ. Health Perspect.* 2011, *119*, 431–438.

[7] Schwarzer, S.; Bono, A. D.; Giuliani, G.; Kluser, S.; Peduzzi, P. *E-waste, the hidden side of IT equipment's manufacturing and use*; United Nations Environment Programme: Geneva, 2005.

[8] StEP *StEP (Solving The E-waste Problem) Annual Report 2010*; United Nations University: Berlin University, 2010.

[9] Osibanjo, O.; Nnorom, I. C., The challenge of electronic waste (e-waste) management in developing countries. *Waste Manag. Res.* 2007, *25*, 489–501.

[10] Ongondo, F. O.; Williams, I. D.; Cherrett, T. J., How are WEEE doing? A global review of the management of electrical and electronic wastes. *Waste Manag.* 2011, *31*, 714–730.

[11] Betts, K., A changing e-waste equation. *Environ. Sci. Technol.* 2010, *44*, 3204.

[12] Sinha-Khetriwal, D.; Kraeuchi, P.; Schwaninger, M., A comparison of electronic waste recycling in Switzerland and in India. *Environ. Impact Asses.* 2005, *25*, 492–504.

[13] Morf, L. S.; Tremp, J.; Gloor, R.; Schuppisser, F.; Stengele, M.; Taverna, R., Metals, non-metals and PCB in electrical and electronic waste – Actual levels in Switzerland. *Waste Manag.* 2007, *27*, 1306–1316.

[14] Duarte, A. T.; Dessuy, M. B.; Silva, M. M.; Vale, M. G. R.; Welz, B., Determination of cadmium and lead in plastic material from waste electronic equipment using solid sampling graphite furnace atomic absorption spectrometry. *Microchem. J.* 2010, *96*, 102–107.

[15] Lim, S.-R.; Schoenung, J. M., Human health and ecological toxicity potentials due to heavy metal content in waste electronic devices with flat panel displays. *J. Hazard. Mater.* 2010, *177*, 251–259.

[16] Wienold, J.; Recknagel, S.; Scharf, H.; Hoppe, M.; Michaelis, M., Elemental analysis of printed circuit boards considering the ROHS regulations. *Waste Manag.* 2011, *31*, 530–535.

[17] Nnorom, I. C.; Osibanjo, O., Toxicity characterization of waste mobile phone plastics. *J. Hazard. Mater.* 2009, *161*, 183–188.

[18] Wu, B. Y.; Chan, Y. C.; Middendorf, A.; Gu, X.; Zhong, H. W., Assessment of toxicity potential of metallic elements in discarded electronics: A case study of mobile phones in China. *J. Environ. Sci.-China* 2008, *20*, 1403–1408.

[19] Wong, M. H.; Wu, S. C.; Deng, W. J.; Yu, X. Z.; Luo, Q.; Leung, A. O. W.; Wong, C. S. C.; Luksemburg, W. J.; Wong, A. S., Export of toxic chemicals – A review of the case of uncontrolled electronic-waste recycling. *Environ. Pollut.* 2007, *149*, 131–140.

[20] Darnerud, P. O., Toxic effects of brominated flame retardants in man and in wildlife. *Environ. Int.* 2003, *29*, 841–853.

[21] Legler, J., New insights into the endocrine disrupting effects of brominated flame retardants. *Chemosphere* 2008, *73*, 216–222.

[22] Wang, Q.; He, A. M.; Gao, B.; Chen, L.; Yu, Q. Z.; Guo, H.; Shi, B. J.; Jiang, P.; Zhang, Z. Y.; Li, P. L.; Sheng, Y. G.; Fu, M. J.; Wu, C. T.; Chen, M. X.; Yuan, J., Increased levels of lead in the blood and frequencies of lymphocytic micronucleated binucleated cells among workers from an electronic-waste recycling site. *J. Environ. Sci. Heal. A* 2011, *46*, 669–676.

[23] Huo, X.; Peng, L.; Xu, X.; Zheng, L.; Qiu, B.; Qi, Z.; Zhang, B.; Han, D.; Piao, Z., Elevated blood lead levels of children in Guiyu, an electronic waste recycling town in China. *Environ. Health Perspect.* 2007, *115*, 1113–1117.

[24] Qin, X.; Xia, X.; Li, Y.; Zhao, Y.; Yang, Z.; Fu, S.; Tian, M.; Zhao, X.; Qin, Z.; Xu, X., Ecotoxicological effects of mixed pollutants resulted from e-wastes recycling and bioaccumulation of polybrominated diphenyl ethers in Chinese loach (*Misgurnus anguillicaudatus*). *J. Environ. Sci.-China* 2009, *21*, 1695–1701.

[25] Tue, N. M.; Sudaryanto, A.; Tu, B. M.; Isobe, T.; Takahashi, S.; Pham, H. V.; Tanabe, S., Accumulation of polychlorinated biphenyls and brominated flame retardants in breast milk from women living in Vietnamese e-waste recycling sites. *Sci. Total Environ.* 2010, *408*, 2155–2162.

[26] Leung, A. O. W.; Chan, J. K. Y.; Xing, G. H.; Xu, Y.; Wu, S. C.; Wong, C. K. C.; Leung, C. K. M.; Wong, M. H., Body burdens of polybrominated diphenyl ethers in childbearing-aged women at an intensive electronic-waste recycling site in China. *Environ. Sci. Pollut. R.* 2010, *17*, 1300–1313.

[27] Toms, L.-M. L.; Hearn, L.; Kennedy, K.; Harden, F.; Bartkow, M.; Temme, C.; Mueller, J. F., Concentrations of polybrominated diphenyl ethers (PBDEs) in matched samples of human milk, dust and indoor air. *Environ. Int.* 2009, *35*, 864–869.

[28] Chatterjee, P., Health costs of recycling. *BMJ* 2008, *337*, 376–377.

[29] Jun-Hui, Z.; Hang, M., Eco-toxicity and metal contamination of paddy soil in an e-wastes recycling area. *J. Hazard. Mater.* 2009, *165*, 744–750.

[30] Robinson, B. H., E-waste: An assessment of global production and environmental impacts. *Sci. Total Environ.* 2009, *408*, 183–191.

[31] Ladou, J.; Lovegrove, S., Export of electronics equipment waste. *Int. J. Occup. Environ. Health* 2008, *14*, 1–10.

[32] Nnorom, I. C.; Osibanjo, O., Overview of electronic waste (e-waste) management practices and legislations, and their poor applications in the developing countries. *Resour. Conserv. Recycl.* 2008, *52*, 843–858.

[33] Rosenfeld, P. E.; Feng, L. G. H., The Export of Hazardous Waste. In *Risks of Hazardous Wastes*, William Andrew Publishing: Boston, 2011; pp 169–184.

[34] Schluep, M.; Hagelueken, C.; Kuehr, R.; Magalini, F.; Maurer, C.; Meskers, C.; Muellera, E.; Wang, F. *Recycling - from E-waste to Resources*; United Nations Environment Programme: Berlin, 2009.

[35] Barba-Gutierrez, Y.; Adenso-Diaz, B.; Hopp, M., An analysis of some environmental consequences of European electrical and electronic waste regulation. *Resour. Conserv. Recycl.* 2008, *52*, 481–495.

[36] Tange, L.; Drohmann, D., Waste electrical and electronic equipment plastics with brominated flame retardants - from legislation to separate treatment - thermal processes. *Polym. Degrad. Stab.* 2005, *88*, 35–40.

[37] Sato, Y.; Oki, M.; Kondo, A.; Takenaka, M.; Satake, H., Rapid analysis of polybrominated diphenyl ethers by ion attachment mass spectrometry. *Anal. Methods-UK* 2010, *2*, 701–706.

[38] Jung, J.; Bae, S.; Lee, L.; Keun Shin, J.; Choi, J.; Lee, S., Rapid identification of brominated flame retardants by using direct exposure probe mass spectrometry. *Microchem. J.* 2009, *91*, 140–146.

[39] Covaci, A.; Voorspoels, S.; de Boer, J., Determination of brominated flame retardants, with emphasis on polybrominated diphenyl ethers (PBDEs) in environmental and human samples-A review. *Environ. Int.* 2003, *29*, 735–756.

[40] Lee, C.-H.; Tang, L.-W.; Popuri, S. R., A study on the recycling of scrap integrated circuits by leaching. *Waste Manag. Res.* 2011, *29*, 677–685.

[41] Sato, Y.; Oki, M.; Kondo, A.; Takenaka, M.; Satake, H., Rapid analysis of polybrominated diphenyl ethers by ion attachment mass spectrometry. *Anal. Methods-UK* 2010, *2*, 701–706.

[42] Vilaplana, F.; Karlsson, P.; Ribes-Greus, A.; Ivarsson, P.; Karlsson, S., Analysis of brominated flame retardants in styrenic polymers: Comparison of the extraction efficiency of ultrasonication, microwave-assisted extraction and pressurised liquid extraction. *J. Chromatogr. A* 2008, *1196-1197*, 139–146.

[43] Zhong, Y.; Peng, P. a.; Yu, Z.; Deng, H., Effects of metals on the transformation of hexabromocyclododecane (HBCD) in solvents: Implications for solvent-based recycling of brominated flame retardants. *Chemosphere* 2010, *81*, 72–78.

[44] Tsydenova, O.; Bengtsson, M., Chemical hazards associated with treatment of waste electrical and electronic equipment. *Waste Manag.* 2011, *31*, 45–58.

[45] Nnorom, I. C.; Osibanjo, O., Sound management of brominated flame retarded (BFR) plastics from electronic wastes: State of the art and options in Nigeria. *Resour. Conserv. Recycl.* 2008, *52*, 1362–1372.

[46] Chi, T. D.; Lee, J.-c.; Pandey, B. D.; Yoo, K.; Jeong, J., Bioleaching of gold and copper from waste mobile phone PCBs by using a cyanogenic bacterium. *Miner. Eng.* 2011, *24*, 1219–1222.

[47] Brandl, H.; Faramarzi, M. A., Microbe-metal-interactions for the biotechnological treatment of metal-containing solid waste. *China Part.* 2006, *4*, 93–97.

[48] Kahhat, R.; Kim, J.; Xu, M.; Allenby, B.; Williams, E.; Zhang, P., Exploring e-waste management systems in the United States. *Resour. Conserv. Recycl.* 2008, *52*, 955–964.

[49] *European Parliament legislative resolution of 3 February 2011 on the proposal for a directive of the European Parliament and of the Council on waste electrical and electronic equipment (WEEE)* European Commission: Brussels, 2011.

[50] Chung, S.-S.; Zhang, C., An evaluation of legislative measures on electrical and electronic waste in the People's Republic of China. *Waste Manag.* 2011, *31,* 2638–2646.

[51] *Regulation (EC) No 1907/2006 of the European Parliament and of the Council of 18 December 2006 concerning the Registration, Evaluation, Authorisation and Restriction of Chemicals (REACH), establishing a European Chemicals Agency.*; European Commission: Brussels, 2006.

[52] Herat, S., Sustainable management of electronic waste (e-waste). *Clean-Soil Air Water* 2007, *35*, 305–310.

[53] Selin, H.; VanDeveer, S. D., Raising global standards - Hazardous substances and E-waste management in the European Union. *Environment* 2006, *48*, 6–18.

[54] Shinkuma, T.; Managi, S., On the effectiveness of a license scheme for E-waste recycling: The challenge of China and India. *Environ. Impact Asses.* 2010, *30*, 262–267.

INDEX

B

C

D

E

H

I

M

Q

R

S

T